GOD'S DOODLE

GOD'S DOODLE

The
LIFE and
TIMES
of the
PENIS

TOM HICKMAN

SOFT SKULL PRESS
an imprint of COUNTERPOINT
Berkeley

First published in Great Britain in 2012 by Square Peg

Typeset and designed by carrdesignstudio.com

Library of Congress Cataloging-in-Publication is available.
Hickman, Thomas.
God's doodle : the life and times of the penis /
Thomas Hickman.
pages cm
Includes bibliographical references and index.
ISBN 978-1-59376-525-5 (pbk.)
1. Penis—Humor. I. Title.
PN6231.P344H53 2013
818'.602—dc23
2013017910

SOFT SKULL PRESS
An imprint of COUNTERPOINT
1919 Fifth Street
Berkeley, CA 94710
www.softskull.com
www.counterpointpress.com

Printed in the United States of America
Distributed by Publishers Group West

10 9 8 7 6 5 4 3 2 1

He's got his root in my soul, has that gentleman!
An' sometimes I don' know what ter do wi' him.
Ay, he's got a will of his own, an' it's hard to suit him.
Yet I wouldn't have him killed.

Lady Chatterley's Lover, D.H. Lawrence

PROLOGUE

As one

It is a truth universally acknowledged that a man in possession of a penis will do some of his thinking with it. Physiologically, this is impossible. But as Bill Clinton's one-time lover Gennifer Flowers observed about the reckless presidential dalliance with White House intern Monica Lewinsky, 'He was thinking with his other head' – an observation that both acknowledges the phenomenon and emphasises the penis's ability to override the higher thought processes, despite lacking the 100 million nerve cells that make up the brain's neuronal highway.

Five hundred years ago, Leonardo da Vinci, the outstanding genius of the Renaissance, puzzled over the relationship that exists between a man and his penis, mirror-writing in one of his notebooks:

> [The penis] has dealings with human intelligence and
> sometimes displays an intelligence of its own; where a man
> may desire it to be stimulated, it remains obstinate and
> follows its own course; and sometimes it moves on its own
> without permission or any thought by its owner. Whether

one is awake or asleep, it does what it pleases; often the man is asleep and it is awake; often the man is awake and it is asleep; or the man would like it to be in action but it refuses; often it desires action and the man forbids it. That is why it seems that this creature often has a life and an intelligence separate from that of the man.

According to the Athenian tragedian Sophocles, to possess a penis is to be 'chained to a madman' – and the madman is capable of seizing control of the possessor's centre of command. *Ven der putz shteht, light der sechel in drerd*, runs the Yiddish proverb: When the prick stands up, the brains get buried in the ground. At such moments, the Japanese say, the possessor is possessed, *sukebe* – a comic fellow dragged along by the mischievous lecher between his legs.

What cannot be denied is that in some respects penis-possession gives its possessor a monocular understanding of the world. From infanthood he comes to consider his penis as an entertainment centre both for himself and for others, the genie's lamp which when rubbed fulfils his wishes (he at least wishes).

He would, of course, indignantly deny that his possession is independent of his psychology and personality: he is, after all, more than the sum of his private parts. But a penis – always – has the potential to inflict humiliation or introduce ethical dilemma. And does the man sport the penis or the penis the man or, to go further, is there truth in the playwright Joe Orton's mischievous assertion that 'a man is nothing more than a life-support system for his penis'? It is the basis for a lifetime's schizophrenia.

Throughout history, women's attitude to the penis has been no less ambivalent than men's. That the penis is capable of coming alive apparently outside the control of its possessor makes women also regard it, on occasion, as an entity in some sense separate from him, which is why, as Simone de Beauvoir observed in the first post-war feminist tract *The Second Sex*, mothers speak to the infant male of his penis as 'a small person ... an alter ego usually more sly ... and more clever than the individual', compounding, from the beginning of his life, his belief that he and it are a duality, like Batman and Robin. Later, more aggressive feminists than de Beauvoir, while heaping opprobrium on the penis's head, haven't shaken free of this double vision. 'You never meet a man alone,' one feminist wrote. 'There are always two of them: him and his penis', the aggrieved tone rather suggesting that the penis should be unstrappable, like a six-shooter in a Wild West saloon handed to the barkeep to ensure it causes no trouble. Feminism has alleged that the 'mere' possession of a penis has been responsible for thousands of years of male domination of religion and philosophy, of political, social and economic thought, of history itself.

Such a claim, one might observe, is over at least one of the heads of the penis-possessor.

PART ONE

**MEASURE
FOR
MEASURE**

Penis size is not really important. Like they say, it's not the size of the boat, it's the length of the mast divided by the surface area of the mainsail and subtracted from the circumference of the bilge pump. Or something like that.

Donna Untrael

'AS INDIVIDUAL AS FACES'

In 1963 the fate of the British government hung on a Cabinet minister's genitals.

Harold Macmillan's Conservative administration was already tottering, having just lost War Minister John Profumo over his affair with the prostitute Christine Keeler, when the 11th Duke of Argyll began divorce proceedings against his wife Margaret, alleging adultery with eighty-eight unnamed men including three royals, three Hollywood actors and not one but two Cabinet ministers. Sensationally, the duke produced Polaroid photographs, then a novelty, one showing his wife, wearing only a pearl necklace, fellating a man in the bathroom of her Belgravia home, and a set of four others showing a man lying on her bed masturbating, which were captioned 'before', 'thinking of you', 'during – oh', and 'finished.' Who was 'the headless man', as the newspapers dubbed him – unidentifiable in the first Polaroid because the camera had cut him off at the neck and in the set because of the angle at which they were taken? A number of names were bandied about, but quickly the actor Douglas Fairbanks Junior and Duncan Sandys, the Minister of Defence who was also Winston Churchill's son-in-law, found themselves in the frame, head to head, so to speak.

Sandys told Macmillan he wasn't the culprit. But if yet another minister was implicated in a sexual scandal the government almost certainly would be toppled and the prime minister wanted to be sure. He therefore instructed Lord Denning, the Master of the Rolls, to investigate. Britain's senior law lord summoned the five most likely suspects, Sandys and Fairbanks among them, to the Treasury, where each had to sign the visitors' book, and got a graphologist to compare their handwriting with the Polaroid photo captions. While he awaited confirmation, Denning had Sandys visit a Harley Street specialist who confirmed that the ministerial genitals were not those in the masturbatory sequence.

In the event, Denning was able to tell the prime minister that the handwriting wasn't Sandys' either, but Fairbanks' (something not made public for nearly forty years). For her part, never once in the rest of her long life did the Duchess of Argyll confirm anything. But she did drop heavy hints that two men, not one, were in the photos: the masturbator was not only not the fellatee he was, in fact, Sandys. It may be superfluous to point out that the duchess's enthusiastic bathroom ministrations made a comparison of one penis with the other impossible.

When in dismissive mood, some women are inclined to say of penises that once you've seen one you've seen them all, but penises are infinite in their variety in size (sic), shape and colouration. Penises can be long, short, fat, thin, stumpy, straight, bulbous or so conical as to get pinched in the tip of a condom, swerved left or right or up or down, circumcised or not, smooth or as wrinkled as a Shar Pei pup; and they present in rosy pink, caramel, peach, lavender, plain chocolate or gunmetal black, largely dependent on the ethnic origin of the possessor, but not entirely: most penises are of a darker hue than possessors' bodies, some startlingly so – 'more suntanned', as the Danish couple Inge and Sten Hegeler delicately put it in *An ABZ of Love*, published in 1963, the same year of 'the headless man', and the biggest-selling sex manual of the time. Significantly, the Hegelers considered penises to be 'as individual as faces'. More significantly, in the following decade, Alex Comfort, writing in *The Joy of Sex*, the biggest-selling sex manual of all time, decreed that penises are also endowed with 'a personality'.

Whether penises are as individual as faces or not, in cartoonish fashion the male genital compendium *in toto* has been likened to a face: the face of a very old man with a particularly ugly nose and an egg tucked into each sagging jowl (undeniably, beyond puberty, every penis looks older than its possessor). Even the head or glans of the penis has been likened to a face, which somewhat stretches the imagination, though perhaps one can see it as 'early foetal'. 'Such a serious little face,' Thelma says of Harry Angstrom's penis (*Rabbit Is Rich*, John Updike) in a moment's rest from fellating him, completing the analogy by noting that his uncircumcised foreskin around the swollen head of his penis is like 'a little bonnet'. Serious, or more especially sad, penises were so intolerable to the poet Bonnie Roberts, she revealed in her poem 'Portrait of a Former Penis Bigot', that she dotted happy faces on her lovers' with a felt-tip pen; a Smiley badge may never look the same again.

Down the centuries the urethral opening at the penis's tip has been referred to as an eye ('Jap-eyed' in modern vernacular, though classical scholars in the 1920s grandly referred to the penis as Polyphemus, the one-eyed Cyclops, son of Poseidon, who was blinded by Odysseus) or said to resemble a tiny mouth, and it to this that the Elizabethan poet Richard Barnfield referred in a sonnet that begins 'sweet coral lips, where nature's treasure lies' (it helps to know that Barnfield was homosexual or one might be misled). In Elizabethan England 'nose' was popular slang for penis and indeed today, on the other side of the world in Japan, the penis is colloquially referred to by the name of a folklore goblin, Tenggu, unfortunate enough to have an outsize olfactory organ!'*

For centuries men have given penises men's names, a matter, you might say, of putting names to faces: in England most popularly Peter, Percy, Rupert and Roger – traditionally a name given to stud bulls and rams – and, still current, John or John-Thomas (now more usually, thanks to D.H. Lawrence's usage in *Lady Chatterley's Lover*) and Willy (a foreshortening of William). But interestingly Dick, although as old as any of these, only joined the penile fraternity in the late nineteenth century and then not because it rhymed with prick, but as a shortened form of dickory dock, cockney rhyming slang for cock. Roger is no longer extant as a nickname (one hesitates to say diminutive) but for centuries has been a well-loved verb – the diary of William Byrd of Westover contains the earliest recording: on 26 December 1711 he wrote of his wife 'I rogered her lustily', and again on 1 January 1712, 'I lay abed till 9 o'clock this morning . . . and rogered her by way of reconciliation'.

Some men give their penises nicknames (can Clinton really have called his Willard?) because, as the joke goes, they don't want to be bossed around by somebody they don't know.

Anglo-Saxon men did not have penises. They were tarse men. Over five hundred or so years, men became pintle men or pillicock men. When these terms from the Middle Ages were considered vulgar in the late sixteenth century, pillicock became shortened to cock (pillicock leaving us with the mildly offensive pillock) and it and prick became the acceptable

---

* See Part One Notes page 60

referents, however surprising to modern ears: in the sixteenth and seventeenth centuries, maids routinely referred fondly to their boyfriend as 'my prick'. By the end of the seventeenth century, prick dropped out of polite society, as did cock, with wider linguistic consequences: apricocks, haycocks and weathercocks became apricots, haystacks and weathervanes, as in the America of the Puritan Fathers water cocks became faucets and cockerels, roosters. Men now sported the yard – derived from a medieval term for rod or staff carried as an indication of authority, not an optimistic measurement of length.

When the eighteenth century developed a liking for Latinate terms, yard finally became penis, and tarse, which had hung on at least in literary circles, now bowed out, much to the regret of scatological poets (penis does not rhyme with arse). The classical Roman term for penis was *mentula*, which one might think had a certain resonance equating as it does to 'little mind'. But eighteenth-century wordsmiths preferred the idiomatic penis, meaning tail, not just to *mentula* but to the most popular Roman slang of *gladius*, or sword – which, as vagina meant sheath or scabbard, fitted nicely.[2]

Glans, the Latin word for the head of the penis erect (meaning acorn, which it resembles, somewhat fancifully), was also adopted into standard English – though most people conversationally stuck to the centuries-old knob, helmet, bell-end and, of course, head. ('Come, Kate, thou art perfect in lying down: come, quick, quick, that I may lay my head in thy lap' – *Henry IV* Part I, III, i, 226–8.)

The rest of the compendium also underwent Latinisation. What the Anglo-Saxons and all those who followed them referred to as cullions, ballocks (later spelt bollocks) or stones (used consistently in the King James Bible of 1611), and from the sixteenth to the eighteenth century as cods (from codpiece), were henceforward testicles, from *testiculus*, witness – Romans thought of their testicles as 'little witnesses of virility' etymologists conclude (see Part 2, 'From Bit Player to Lead').

In centuries intimately conversant with the Bible, Adam's arsenal, Nimrod (the mighty hunter) and Aaron's rod (the patriarch's staff, which blossomed and yielded almonds) were among the sobriquets, neologisms and tropes men devised for their genitals – not now likely to be found on Internet sites compendiously devoted to genital terminology. Down the ages men have also applied every synonym imaginable to the penis: assorted vegetables and fruits, small animals or animal parts, and reptiles – the snake and eel are constants in almost every culture, as is the phallic-headed snapping turtle in the cultures of the Middle East. The Italians still refer colloquially to the penis as a bird or fish, just like the Sumerians, the world's first civilisation, over five and a half thousand years ago. Specific weapons and tools have always loomed large in the penile vocabulary, 'sword' as popular elsewhere as it was in Ancient Rome – Shakespeare employed it as well as such terms as pike, lance, pistol and poll-axe. As weaponry advanced the penis equated to ever-more potent munitions including, in recent years, torpedo, bazooka and rocket.

But all of this verbal ingenuity aside, cock, prick – and the generic tool and weapon – remain the words most commonly used for the penis in English, as do their equivalents in other languages, with balls and nuts (a foreshortening of the seventeenth-century coinage nutmegs) for the attendant testicles. The British continue to have a fondness for bollocks, knackers (a verb in the Middle Ages meaning to geld, perhaps not the happiest association), cobblers (more cockney rhyming slang, from cobbler's awls) and, harking back to colonial days in India, goolies (from a Hindi word for any round object). The Americans' favoured alternative for testicles is rocks, stones being, one assumes, not big enough in a country where everything must be bigger.

Just as some believe in face reading, phrenology, palmistry or podomancy (foot reading), there are some who believe that the complementary study of phallomancy, which has a long tradition in India and Tibet, can divine both a man's character and fortune. Tibetans believe it's unlucky for a man to be over-endowed: should his penis reach the bottom of his heels while squatting, his life will be full of sorrow; should, however, his penis be no more than six finger widths' long he will be rich and a good husband. Hindus have similar beliefs, expounded in the *Brihat Samhita*, a Sanskrit astrological treatise written in the sixth century AD. The over-endowed man will be poor and without sons; the man whose penis is straight, small and sinewy will be rich, as will the man the head of whose penis is not very large. The man whose penis inclines towards the left is another who will always know poverty, as will the man the head of whose penis is depressed in the middle – and this man will father only daughters. According to the *Brihat Samhita*, the man who has perfectly matched testicles will be a king. The compensation for men with mismatched testicles is that they will be fond of sex. The question for non-Tibetan and non-Hindu males who have mismatched testicles and a fondness for sex, is why they think there might be something in phallomancy on the basis of the last reading, yet dismiss the rest as Eastern nonsense.

Matters of size

How big is big? How small is small? What is average? Where do you fit in? Sixteen hundred years ago, when Vatsyayana compiled the *Kama Sutra*, the world's oldest sex manual, drawing on texts that were already up to 800 years old, he classified men according to the size of their erect lingam (penis). Hares were equivalent to the width of six fingers, bulls to eight and stallions to twelve, spans of between 4.5 inches and 9 inches, or 6 and 12, depending on the size of the hand – a detail that Vatsyayana omitted, though most Asians being small-boned, have small hands.

Such imprecision was not for the Victorians. They were not the first to attempt scientific scrutiny of human sexuality, but they were the first to attempt it on a statistical and empirical basis, driven by expanding knowledge and the new discipline of psychoanalysis. Not unnaturally, the penis, and the size of the penis, principally erect, was central to this study. Dr Robert Letou Dickinson spent a lifetime making hundreds of drawings from life showing penises in repose and arousal (which he published as the *Atlas of Human Sex Anatomy* only in 1949, when he was eighty-eight). One erection he included was 13.5 inches in length and 6.25 in circumference, the largest ever medically verified. In recent years a New York clubber, Jonah Falcon, has shown enough journalists that his is the equal in both dimensions for there to be little doubt, medically verified or not.

Erotic fiction abounds with penises of such stature. In *Fanny Hill*, the most widely known erotic novel in English (which John Cleland wrote 250 years ago, to get him out of debtors' prison), the eponymous heroine encounters organs 'not less than my wrist and at least three of my handfuls long'; 'a maypole of so enormous standard that, had proportions been observed, it must have belonged to a young giant'; and, most impressively, one whose 'enormous head seemed, in hue and size, not unlike a common sheep's heart: then, you might have rolled dice along the broad back of the body of it'. But the vast majority of those appended to men are lesser things. After the Second World War, Alfred Kinsey conducted 1,800 exhaustive interviews with men and amassed penile data on a total of 3,500 before stating in *Sexual Behavior in the Human Male*, published in 1948, that the average erect penis was 6.2 inches, with 'most individuals' in the 4.8 to 8.5 inch range and only 'extreme cases that are both longer and shorter'. In fact, the shortest erection Kinsey encountered was 1 inch, and the longest 10.5 inches. The erection with the least circumference was 2.25 inches, the greatest, over 8, the average 4.75.

Kinsey was a professor of zoology at Indiana University, with a worldwide reputation for his study of gall wasps. It was only after the university set up a course on sexuality in matrimony and asked him to teach it that he turned to the investigation of sex and eventually founded his famous institute. One female student was made so enthusiastic by his slides and graphic descriptions that she wrote, 'To me, the behavior of the penis was already awe-inspiring; now it seems even more wonderful.' Another female student was evidently less enthusiastic. When one day Kinsey broke off from a lecture and asked her which human organ was capable of the greatest expansion, she flushed. 'Professor Kinsey, you have no right to ask me that question,' she said. Kinsey replied, 'I was thinking of the eye – the iris of the eye. And you, young lady, are in for a great disappointment.'

Despite Kinsey's voluminous data-gathering, there was still no accurate information about the physiology of sex, until the husband and wife team of William Masters and Virginia Johnson, following in his wake, conducted eleven years of empirical research. Kinsey, for the most part, had extrapolated his findings from questionnaires. In the more permissive 1960s, Masters and Johnson attached electrodes to some seven hundred men and women and filmed and monitored them engaged in sexual activity. But while they basically confirmed Kinsey's findings about the dimensions of the penis, Masters and Johnson also made men with smaller penises – smaller, that is, when flaccid – feel good about themselves, because they noted something that Kinsey did not record: that the smaller the organ the greater the proportional increase during erection.

The average flaccid penis, Masters and Johnson said, was between 3 and 5 inches. In their research they compared a group of men at the lower end of this range with a group of those at the higher, and while the latter on erection had increased slightly less than 3 inches (one 4.5-er added only 2 inches), the former nearly doubled (one 3-incher added 3.33 inches). One participant who when flaccid showed no penile shaft whatsoever – the kind of penis that Fanny Hill described as 'scarce showing its tip above the sprout of hairy curls that clothed those parts, as you may have seen a wren peep his head out of the grass' – 'grew to normal proportions'. Masters and Johnson's important conclusion was that erection is 'the great equaliser'; unaroused penises vary considerably, but there is a tendency for things to even out somewhat when they go to red alert.

There is, to put this another way, no correlation between sizes of flaccid and erect penises, just as there is no correlation between the erection and the bodily frame, which Masters and Johnson also demonstrated, and none between the erection and the size of the hands, feet or nose, as others have demonstrated, although popular myth continues to maintain otherwise, sometimes in inverse proportion. It's true that the Hox gene, which controls initial growth of the genitals of male (and female) foetuses, also controls that of the hands and feet, but the size and shape of hands, feet and genitals are ultimately determined by many genes. A big-framed man can have a big nose, big feet and hands like a bare-knuckle bruiser and still possess a small penis. There's a weak correlation between penis erectile length and girth, but men are nearly as likely to have a (relatively) thin penis as a (relatively) bulky one and any combination of length and girth, with the qualification that the penis of exceptional length is rarely of exceptional circumference. At the risk of stating the obvious, genital size, like all genetic traits, is hereditary, if not necessarily so. There is every biological reason for believing the father of actor Ewan McGregor who, after his son's impressive appendage was lingeringly displayed in the film *The Pillow Book*, sent him a fax reading: 'Glad to see you have inherited one of my major attributes.'

Racial assortments

Kinsey collected his data exclusively from among white Americans. That he didn't include black males was governed by the socio-political climate of his time: post-war America was still a racially segregated country. Had he incorporated African American data he certainly would not have been able to draw racial comparisons from it that might have allowed some interpretation of black ascendancy. Even a quarter of a century later Beth Day, writing *Sexual Life between Blacks and Whites*, was nervous about addressing the issue. Only noting that studies of comparative genital size were few and inconclusive, she went no further than to cite Masters and Johnson's finding about larger penises tending to increase less on erection, concluding: 'Considering this apparent equalisation, the major difference, then, in genital size between black males as a group and white males as a group is psychological.'

The negroid/caucasoid question has arisen throughout time. In the second century BC Galen, physician to three Roman emperors and until the Enlightenment the standard medical authority, wrote that the black man 'has a long penis and great merriment'. Between the seventeenth and nineteenth centuries, Europeans arriving on the African continent were struck by the natives' 'large Propagators', which the French army surgeon and anthropologist Jacob Sutor was convinced were caused by circumcision, the foreskin being, he thought, a kind of compression cap. In 1708 the English surgeon Charles White wrote, 'That the PENIS [his capitalisation] of an African is larger than that of a European has I believe been shewn in every anatomical school in London. Preparations of them are preserved in most anatomical museums, and I have one in mine.' Richard Jobson, treasure-hunting along the Gambia River in West Africa, wrote that the Mandingo tribesmen were 'furnisht with members so big as to be burdensome to them'. Others recorded 'terrific machines' of as much as 12 inches in length, the kind of measurement that at the beginning of the twentieth century made the homosexual British consular official Sir Roger Casement tremble with excitement while in Peru. Casement (who converted to Irish nationalism and was hanged for treason in 1916), wrote in his *Black Diaries*, suppressed until 1956: 'saw the young Peruvian Negro soldier leaving barracks with erection under white knickers – it was halfway to knees! Fully one foot long.'

Evidence of a more clinical nature was published in 1935 by the authoritative *L'Ethnologie du Sens Genitale*, but it wasn't until thirty years after Kinsley ducked the issue of the negroid penis (as did Latou Dickinson in 1949, in his *Atlas of Human Sex Anatomy*, which contained hundreds of drawings of penises but not one of them black) that the institute Kinsey founded, which was and remains the leading authority in its field, felt able to release material on the black/white issue. This – which coincidentally downgraded the non-black erection from 6.2 inches to 6.1 – indicated that the average black counterpart was longer (6.4) and thicker (4.9 against 4.8), and that almost twice as many blacks (13.6 per cent) as whites (7.5 per cent) pushed beyond the 7-inch barrier. But the institute's basic conclusion, which wasn't exactly a surprise, could hardly be termed definitive: while by this time (1979) it had 10,000 men on its database, only 400 of them were black. Understandably, the institute emphasised that comparisons required caution. A decade later, however, and under no politically correct restraint when contributing an article entitled 'Race Difference in Sexual Behavior: Testing an Evolutionary Hypothesis' to the *Journal Research in Personality*, academics John Philippe Rushton and A.F. Bogaert, having averaged ethnographic data from all available sources, concluded that the erections of caucasians were 5.5 to 6 inches in length and 4.7 inches in circumference and of blacks 6.25 to 8 inches (6.2 inches in circumference) – while those of 'orientals' were 4 to 5.5 inches (3.9 inches in circumference). Data from among mixed-blood males in the French West Indies indicated that penis size increased proportionate to the amount of black blood.

For a decade, Rushton and Bogaert's extrapolations were definitive enough for everybody until the Internet made more detailed research possible and two significant on-line surveys were launched in the 1990s – one by the makers of Durex condoms (whose interest was condom fit and the incidence of slippage and breakage), the other, the Definitive Penis Survey (which despite its frisky title was acknowledged by Durex as a serious research source), by medical researcher Richard Edwards. Both anticipated analysing the information they gathered by ethnicity. Neither was able: too few of the 3,000-plus participants each website attracted were non-white. The Definitive Penis Survey, however, offered 'tentative' conclusions. Extraordinarily, one was that average *white* erections are longer (6.5 inches) than black (6.1 inches). Offsetting that, however, was the observation that where Kinsey's all-white data indicated that three men in a hundred have an erection less than 5 inches, the number of black males in this category is *statistically* immeasurable although, undoubtedly, they exist: the law of averages cannot be gainsaid.

'The dozen or so jungle bunnies I have trafficked with were perfectly ordinary in that department . . . in fact, two were hung like chipmunks', comments Gore Vidal's vitriolic – and racist – transsexual protagonist in *Myra Breckinridge*, though to suggest that in a random sample of a dozen black males none would be bigger than average and two would be smaller is statistically nonsensical. Of course, many black males are statistically average and the British TV chef Ainsley Harriott is one who was happy to say so. A decade or so ago, after he'd done a *Full Monty* strip for the Children in Need charity, a journalist observed that he was not exactly Lynford Christie in the lunchbox department (see Part 2, 'Accessorise – or Aggrandise?'). To which Harriott retorted amiably: 'I'm not twenty-eight with a six-pack. I'm forty-one with two kids. But I like to think my chest isn't that bad.'

What is not disputed in any major survey is that the black penis is observably more visible in the flaccid state than the white – the Kinsey Institute gives figures of 4.3 inches long and 3.7 inches round, against 3.8 and 3.1. A theory about this relative inequality is that, while the penises of men from colder climates spend more time drawn closer to the body's heat, penises of men in warm climates simply hang – 'as long as whips', thinks Harry Angstrom, the eponymous hero of John Updike's quartet of *Rabbit* novels (the black penis gaining disproportionately from the simile). The hot–cold theory appears to be borne out by a research project that equates African American males more closely with the statistical American norm than with males from the Caribbean. Against it, however, is that Asians who come from warm climates are not beneficiaries of it.

Statistics show that the penises of those from the Far East, South East Asia and the Indian subcontinent are smaller than the world median; and according to available data and much anecdotal evidence, a large Asian penis is exceptional.

Investigating Asian penis size, an Asian Los Angeles writer, L.T. Goto, found a Japanese American who hoisted a 7-inch erection, something so ethnically rare that it had gained him 'instant notoriety after dating someone in the Los Angeles Asian American community'. The Definitive Penis Survey (which appears to have been moribund since 2002) had difficulty attracting Asian interest, rather indicating that Asians have better things to do than measure their appendages. The one ethnographic observation that the Durex survey has allowed itself to make is that the erect penises of males in the Far East are up to 20 mm (just over 0.75 of an inch) less in both length and girth than whites – although other survey material substantively indicates that this is a little generous to Indo-Chinese.[3]

Where most people are prepared to see racial variations in penis size as just another racial diversity, there are some who deny that such variations exist, on racist grounds. The topic was given fierce airing in the mid-1990s with the publication of Rushton's *Race, Evolution and Behavior*. In this work of speculative biology, Rushton, a professor at the University of Western Ontario and a Guggenheim Fellow with two doctorates from the University of London, used over sixty variables in a comparative study of Asians, whites and blacks. By all of these, including brain size and intelligence, he deduced that the three groups always rank in that order. It was not this that got him vilified: it was his conclusion that there is a correlation between the size of the organs of generation and cogitation – that Asians have small genitals and high intelligence, that blacks are their opposite and, as by every other comparative measure, that whites occupy the middle ground. Most critics decided that Rushton had been sidetracked by an aberration, concluding that the socio-biological value of a big brain and a small penis is no clearer than that of the alternative arrangement, however popular, given a choice, that would be among at least some males everywhere – including, you can be sure, many Asians.

In the decades since his death, Kinsey's institute has continued to amass and correlate penile information extrapolated from many sources, including other research in which it engages (most recently the psycho-physiological sexual response in men) and the work of urologists involved in the relatively new area of augmentation phalloplasty (see Part 3, 'Desperately seeking solutions'). It has backed away from the question of ethnicity. The institute website, sensibly one might think, does not give a single figure for the average worldwide erection, preferring to declare that this is between 5 and 7 inches, with a circumference of 4 to 6 inches – the founding father's findings have not been invalidated by black highs and Asian lows.

Size Matters?

The importance of penis size is sowed early in the male mind. When a very small boy comes face-to-face (perhaps literally) with the adult penis, he is disbelieving: it cannot be that his own small tag of flesh bears any relation to something that appears to have more in common with the Gruffalo. Does such an encounter, wondered Alexander Waugh (*Fathers and Sons*), 'puncture or augment the sexual confidence of young males?' When he caught his young son standing on a bucket outside the window to catch a glimpse of his '*zones privés*' Waugh composed a verse that he made his son memorise, reading in part:

> For he is but a craven fool
> Who muses 'pon his father's tool,
> Or creeps and peeps and tries to spy
> What lies within poor Papa's fly.

But a boy needs to *know*. And whether he finds out by accident or design, when he establishes that one day his, too, will look like this, he can hardly wait, like Portnoy, who wants to exchange his 'fingertip of a penis' for something the equal of his father's 'schlong', which

> brings to mind the fire hoses coiled along the corridors at school. Schlong: the word somehow catches exactly the brutishness, the meatishness, that I admire so, the sheer mindless, weighty and unselfconscious dangle of that living piece of hose. (*Portnoy's Complaint*, Philip Roth)

The worry that his penis will not make the transformation into the alpha version is excruciating to most males in early, hormonally concussed pubescence, a state neatly caught in this confessional piece by a contributor to *Cosmopolitan* magazine:

> The year was 1984. I was 12. One day, a few of my fellow 12-year-olds and I were in the changing room when an older and shall-remain-nameless rugby player strolled in. He stripped and was walking to the shower when he noticed several pubescent boys staring at him, flushed and slack-jawed, upon which he turned in our direction and announced: 'What's the matter, boys? You never seen 18 inches of swaying death before?' Let's just say, there's not a woman alive who can make me feel *that* small.

The outcome is that few males when grown to man's estate free themselves entirely from some preoccupation with penis size, which, Alex Comfort noted in *The Joy of Sex*, is 'built-in biologically' and labelled by the anthropologist Jared Diamond, when professor of physiology at the University of California, as an 'obsession'. Yet it appears that whereas for centuries this has played out as meaning 'big', among the Ancient Greeks the reverse was the case. Penile taste in Athens ran to the small and taut (in the plays of Aristophanes, diminutives such as *posthion*, little prick, were terms of endearment) with big penises considered coarse and ugly and only possessed (the Athenians said) by barbarians and, in their own mythology, satyrs, which were comic in a slapstick way, though we derive our more subtle 'satire' from the word. 'Our modern stag party jokes of well-endowed men', wrote Eva Keuls (*Reign of the Phallus*) 'would have been lost on the Athenians'. The Romans, however, thought little differently to modern man; they had a predilection for big penises – Roman generals sometimes promoted men on the generosity of their genital addenda. 'If from the baths you hear a round of applause,' wrote the creator of the epigram, Marcus Martialis (known in English as Martial), 'Maron's great prick is bound to be the cause.' And like modern man, the Romans heaped ridicule on the head (sic) of anyone plainly below the average, as did the poet Catullus of a fellow Roman 'whose little dagger, hanging more limply than the tender beet, never raised itself to the middle of his tunic' – mockery that 2,000 years later finds an echo in a magazine article by a journalist sent to write a piece on a nudist camp, who spied 'a shy little thing, a nouvelle cuisine portion of garnished mollusc', which made him think that if he were the possessor 'I would be at home in a darkened bedroom, battling with the weights and measures'.

Covertly, throughout time, men have competed with other men with their penises. If their distant ancestors beat their penises against their abdomens to discourage competitors or shook them in the face of inferiors, as some primate species continue to do, men have maintained the practice, if only figuratively – in Japanese art, which is characterised by immense exaggeration of the sexual organs, men are often seen duelling with them. The American politician Walter Mondale once said after some political disagreement with Bush senior, 'George Bush doesn't have the manhood to apologise', to which Bush felt compelled to retort, 'Well, on the manhood thing, I'll put mine up against his anytime' – sounding rather like Grumio in *The Taming of the Shrew* who responds to Curtis's insult about his appendage by saying, 'Am I but three inches? Why, thy horn is a foot; and so long am I at the least.' No man wants to be taken for what Shakespeare termed a 'three-inch fool', a truth not lost on the Tel Aviv advertising agency which in 1994 devised a poster campaign aimed at the city's notoriously bad drivers: 'Research proves aggressive drivers have small penises.' No wonder so many penis-possessors have a compulsion to give their penis a surreptitious tug in the changing room when other men are around – as Alan Bates and Oliver Reed admitted they did before filming the nude wrestling scene in *Women in Love* – to make sure it looks its competitive best.

All of which leads to what Rosalind Miles describes in *The Rites of Man: Love, Sex and Death in the Making of the Male* as a 'lifelong habit of surreptitious cock-watching. In public lavatories, swimming baths, gyms, even at the ballet, [men] will always check out the opposition.'

True or false? In one of its many sexual surveys, *Cosmopolitan* magazine asked men if 'they secretly check out another man's equipment' when standing at a urinal in a public toilet; it also asked if they would prefer to have a 3-inch penis and earn £100,000 a year or have a 10-inch penis and earn £10,000. The survey was, of course, essentially frivolous and without doubt attracted its share of frivolous answers. Nonetheless, the response was revealing. To question one, 82 per cent of respondents said 'never', 16 per cent said 'sometimes' and 2 per cent said 'always', whereas in practice the reverse order might be more likely – the instinct to take a surreptitious look is so powerful that many men do it without realising; and many who don't, consciously resist their instinct only because it breaches social etiquette – other than, perhaps, in gay clubs.

> Waiting in the ringing brightness of the lavatory, he felt
> a tinge of loneliness, and wondered where Danny was.
> Everyone was busy here, men in pairs queuing for the
> lock-ups, others in shorts or torn jeans nodding tightly
> to the music, caught in their accelerating inner worlds.
> A guy in fatigues half-turned and beckoned him over to
> share his stall – Alex leant on his shoulder and looked
> down at his big curved dick peeing in intermittent spurts.
> He unbuttoned and slid in his hand and…there it was, so
> shrivelled that he shielded it from his friend, who said,
> 'You're all right, you're off your face,' and, 'You can do it,'
> and then, hungrily, 'Well, give us a look,' while he stroked
> himself and stared and stared. (*The Spell*, Alan Hollinghurst)

'Almost every male seems to envy someone else's penis,' wrote Dr Bernie Zilbergeld in *The New Male Sexuality*. Woody Allen had already got there in the 1983 film *Zelig* with the line: 'I worked with Freud in Vienna. We broke over the concept of penis envy. He thought it should be limited to women.'

To return to the *Cosmo* question about penis size and money, which like the other nudged respondents towards the 'appropriate' reply (if, that is, they wished to appear mature), 42 per cent said 'yes' to the 'inappropriate' second option – a powerful indication, one might surmise, of how irresistible is men's desire for a greater-than-average penis.

Is it any wonder that *Forum*, the international journal of sexual relations, thinks it has 'probably printed more words about penises than any other part of the body, male or female' (no surprise that a number of its articles on the subject over the years have been headlined 'Man's Best Friend'); or that the most frequent question on all Internet Q&A sex sites continues to be, Is size important? A downloadable chart of four outline drawings ('low average' to 'extraordinarily large') can be found on the net, which a man can print out and use as a template against which to judge himself. There should be solace for the average man in knowing that he is statistically within touching distance, as it were, of some 90 per cent of all his fellows. But even the most balanced of men is capable of half-believing that he is under-endowed – the Hunt survey for *Playboy* in the 1970s indicated that more than two-thirds of men thought that 'something more in their shorts would make a difference'; and 'the majority' of the 7,000 men interviewed by Shere Hite for *The Hite Report on Male Sexuality* 'wished they were bigger'. Many American men, according to the Kinsey Institute, believe the average erection is 10 inches – this despite (or because of) frequently accessing Internet pornography in which participants have shaved off their pubic hair to increase visibility and many have used a vascular device to pump up temporarily.

If few men would say no to having a bigger penis, most are not so discontented with what they have that they seek the means of achieving it. There have always been supposed ways and means. The world's oldest sexual guide, the *Kama Sutra*, counselled men to apply the hairs of a poisonous tree insect for ten nights and sleep face downwards on a wooden bed, 'letting one's sex hang through a hole' – the swelling was supposed to last a lifetime. Today, the siren voice of the Internet offers a multitude of preparations. In almost every time and culture some men have hung weights or connected stretching devices; and websites offer everything from a simple elastic band that fits around the corona or crown and fastens to another band around the thigh, to elaborate contraptions of rings and rods, springs and tension bars. Other sites promote the centuries-old Sudanese Arab technique of *jelq*, a daily technique of stroking the penis to erection from base to tip but stopping short of ejaculation, and then starting the process again and again (rich families used to send their sons to a *mehbil* or athletic club for an attendant to save them the drudgery of doing it themselves). 'Gain three inches' urge emails for creams and patches that drop unsolicited into millions of mailboxes daily.

Since the early 1990s, plastic surgery has been a new enticement. A penis can be thickened by implanting strips of fat taken from other parts of the body (usually buttocks or love handles) under the penile skin; and it can be lengthened, by cutting the ligament that anchors it to the pubic area. But phalloplasty is still considered to be in the early stages and professional associations still distance themselves from it because the results can be less than satisfactory. The body naturally reabsorbs fat and when this happens what remains unabsorbed can create lumps and bumps; even if this doesn't happen, many men find that the transplanted fat makes their erection feel as if it were wearing a padded jacket. And a penis lengthened by detaching it from the suspensory ligament may appear to arise from the scrotum rather than the abdominal wall – and be lucky to rise even half-mast; indeed, some erections after surgery point to the ground. In the worst of outcomes, some men have reported, on erection their penis flaps around like a running hose left on the ground. There can be deformity and pain – and post-surgery scarring has been known to cause a penis to retract, becoming smaller than it was before, a situation so harrowing that in one English court action a lawyer likened his client's emotional state to post-traumatic stress disorder.

There was considerable worry in 1993 for thousands of men who had had penis-enlargement operations in Thailand, after the *Bangkok Post* reported that unscrupulous surgeons had been injecting a mixture of olive oil, chalk and other ingredients into their clients' penises to achieve the required result; the Chiang Mai hospital had even seen penises containing portions of the Bangkok telephone directory. More than a few men, particularly in South East Asia and Japan but also in the US and UK, have taken a DIY approach to an instant increase in size by injecting Vaseline, paraffin and other oils – with calamitous results: serious infections, gangrene leading in some cases to amputation, or erectile dysfunction. In 2002 a thirty-one-year-old showed up at the Institute of Urology and Nephrology in London for treatment for gross abnormality and ulceration after using a high-pressure pneumatic grease gun.

Yet despite the potential hazards, some tens of thousands of men around the world have submitted themselves to the operating table. That nine in ten were already of average or greater dimensions, and psychological profiling indicated that very few had a clinical psychological need, speaks volumes about the significance some men attach to size. Like women who have breast implants, their need is to impress – but to impress other men rather than women, even though the increase is likely to be only an inch at most in the flaccid state and nothing discernible in the erect one. And like women who return for even bigger implants, there are men who return for even more enlargement. One surgeon gave up penile cosmetic practice after a new patient, who'd already had four operations elsewhere, came to him for a fifth, presenting for examination a penis as wide as it was long – 'a dick', the surgeon said, 'like a beer can'.

According to Masters and Johnson, so insecure are men on the subject of their penises that most at some time in their life are likely to suffer from feelings of inadequacy – a size syndrome known as body dysmorphia, common among bodybuilders. Like the anorexic, the dysmorphic looks in the mirror and sees a distorted image: a body or body part that is not as it is. In the throes of penile dysmorphia men are quite capable of believing that the 18-inch appendage once claimed by the porno magazine model Long Dong Silver, or 'Mr Torpedo', and the 15-inch one sported by Dick Diggler in the film *Boogie Nights*, are not confections in latex. Scott Fitzgerald was someone temporarily in an extreme state of the condition in Paris in the 1920s when he asked fellow novelist Ernest Hemingway to lunch because he had something important to ask him. Fitzgerald was nervous and avoided addressing what was worrying him. Finally, at the end of the meal, he blurted out that it was 'a matter of measurements'. Zelda, his wife, had said that he could never make any woman happy, because of 'the way I was built'. Fitzgerald, who had not slept with anyone else, didn't know whether what she said was true. As he wrote in *A Moveable Feast*, the Paris notebook published after his death, Hemingway took Fitzgerald to the gents, assured him on examination that he was normal, and advised him to have a look at the statues in the Louvre. Fitzgerald wasn't convinced; the statues might not be accurate. He was not convinced by Hemingway's assurance that 'Most people would settle for them.' He wasn't convinced even after Hemingway dragged him down to the museum to see for himself.

Forty years later, by then an impotent alcoholic whose creative juices had dried up, Hemingway blew his brains out. And how good his reassurances to Fitzgerald were could be questioned after his friend and fellow novelist Sydney Franklin was quoted as saying, 'I've always thought that his problem was that he was worried about his pincha [penis] . . . the size of a 30/30 shell' – about the size of a little finger. The bullfighting, big-game-shooting lifestyle appears to have been over-compensation for under-endowment.

For many of the three men in a hundred who are statistically said to be under-endowed (that is, below the 5–7-inch erectile median) – and the one man in a hundred unfortunate enough to suffer from what the medical profession insensitively terms a 'micropenis', a penis that achieves under 2.5 inches – the feeling of inadequacy, even deep despair, dominates their lives. Prince Camillo Borghese, second husband of Napoleon Bonaparte's dissipated sister Pauline, fled the French court after she dismissed him for being '*si drolement petit*'. The actor Montgomery Clift was not only tortured by his homosexuality but by a penis that earned him the unedifying diminutive 'Princess Tinymeat'. Only someone as compulsively confessional as the impressionist painter Salvador Dalí would make public his lack of size. In *Unspeakable Confessions* he wrote:

> Naked, and comparing myself to my schoolfriends,
> I discovered that my penis was small, pitiful and soft.
> I can recall a pornographic novel whose Don Juan
> machine-gunned female genitals with ferocious glee,
> saying that he enjoyed hearing women creak like
> watermelons. I convinced myself that I would never be
> able to make a woman creak like a watermelon.

Dalí tried to pretend that he did not mind, though the sexually neurotic landscape of his paintings suggests otherwise. His 'anxiety about the small size of his penis was no exaggeration', comments Ian Gibson in *The Shameful Life of Salvador Dalí* – and the anxiety stayed with him always. Is the American TV and radio shock jock Howard Stern another compulsive confessionalist? He certainly can't stop talking about having a small penis and claimed in his autobiography *Private Parts* that he always goes to a stall rather than a urinal: 'God forbid someone should see my puny pecker. I barely clear the zipper.' Perhaps he's just a well-balanced extrovert free of Daliesque neurosis. Or joking (his first wife Alison said his penis was 'fine'). Perhaps the Spanish crooner Enrique Iglesias was joking when he announced that he wanted to launch a range of extra-small condoms for men like himself who were not well endowed and had a problem finding any to fit. After Lifestyle Condoms offered him $1 million to model their wares, the boyfriend of one-time tennis beauty Anna Kournikova said he'd only been joking. If he was and Stern is, few men share their sense of humour, certainly not 'NW, Worcester', a twenty-five-year-old who wrote to *Forum* advice column, telling of the humiliation of daily teasing in his school's communal showers, before continuing:

> For many years I hated being seen naked but have
> managed to overcome this to a certain degree, although
> I do still feel very unhappy with men seeing my penis,
> and my few sexual conquests so far have always involved
> lovemaking in the dark. In fact, my first three lovers
> never saw my penis at all. Eventually I found a really
> understanding lady who has really helped me to come to
> terms with this problem [but] two things still depress me.
> Often I see pre-pubescent boys at the gym and swimming
> pool I attend with bigger penises than mine and I admit I
> feel hugely jealous and inadequate.

Penis envy, or rather penis fear – fear of the imagined size and potency of another male's sexual organ – is at its heightened worst when a woman enters the equation. The eighteenth-century essayist and critic William Hazlitt, futilely obsessed with the promiscuous daughter of his landlord, believed her 'mad for size', agonised about what his rivals had on offer for her favours, and was driven almost to insanity by an overheard exchange about 'the seven inches' of another lodger. His thinly disguised account of the whole affair, *Liber amoris*, was described by Thomas De Quincey as an exorcism, 'an explosion of frenzy . . . to empty his overburdened spirit'. Humbert Humbert, jealously guarding the nymphet Lolita, is in a paroxysm of anxiety at a swimming pool as he watches a bather, whom he does not yet know is Quilty, his rival and nemesis, follow her with his eyes,

> his navel pulsating, his hirsute thighs dripping with bright droplets, his tight wet black bathing trunks bloated and bursting with vigour where his great fat bullybag was pulled up and back like a padded shield over his reversed beasthood.
>
> (*Lolita*, Vladimir Nabokov)

Anxiety becomes agony when a man knows that another has penetrated the woman he loves. Geoffrey Firmin in Malcolm Lowry's *Under the Volcano* is just such a man. Waiting for Jacques Laruelle to finish showering, the sudden impact of the memory that

> that hideously elongated cucumiform bundle of blue nerves and gills below the steaming unselfconscious stomach had sought pleasure in his wife's body brought him trembling to his feet.

At least such knowledge is usually private or known only within a small circle. Not so in Elizabethan England, where a man with an unfaithful wife was often savagely derided, especially if he meekly accepted his situation; called a 'wittol' (cuckold) he had horns or antlers hung on his house, neighbours made horn signs at him with their fingers – and the most easily mocked were paraded around their parish with horns on their head.

LIES, DAMNED LIES AND SELF-MEASUREMENTS

Intellectually, a man knows that the size of his penis shouldn't be specifically relevant in a relationship, to him or to a woman. His common sense tells him that it will certainly not be the major or controlling factor in a woman's response to him. And yet . . . he can't help believing that it is.

When the Kinsey Institute reviewed its founder's data thirty years after it was published, in the light of subsequent findings, it showed that one man in a hundred reaches beyond the 5 to 7-inch erectile median to 8; that seven men in a thousand go beyond 8; and only one in a thousand touches 9. But Durex and the Definitive Penis Internet surveys, while stressing that their core findings are consistent with Kinsey, have cautiously proposed that there are more very big penises – between four and seven men in every hundred reaching 8 inches, between thirty and forty in every thousand reaching 9, and between ten and thirty in a thousand reaching beyond. And where the institute's data showed that erections above 9 inches are so rare (a word, incidentally, that Kinsey himself always used rather than 'big') as to be *statistically* immeasurable, both surveys have suggested that one man in a hundred posts double figures. In the round, the institute found that eighteen men in a thousand have an erection over the median; Durex and the Definitive Penis propose this figure to be between four and eight times greater. Could Kinsey have been so wrong?

The problem for researchers has been that they have had to rely on participants providing their own measurements. The bulk of Kinsey's data came from self-measurements (marked off on the edge of returned postcards); all the data in the Durex and the Definitive Penis survey undertakings were collected in this way – the DPS giving the average erection as 6.3 inches, with Durex giving it as 6.4. Are penises, then, like people, getting bigger? If men's ears have pricked up at this point, the answer is no: the depersonalised and anonymous nature of the Internet almost certainly explains the apparent increase. Not that Durex and the DPS have not taken safeguards against humorists and delusionists. Durex eliminates extreme replies: lengths under 75 mm (3 inches), 'the size of a large chilli', and those over 250 mm (a touch under 10 inches), 'the size of a large cucumber'. The Definitive Penis Survey has disregarded the blatantly fraudulent ('17-year-old lawyers and those claiming American Zulu warrior ancestry') and eliminates the bottom 1 per cent and the top 2 per cent of replies; additionally the website has asked participants to provide an electronically transmitted photo which includes a tape measure.

Averaging the averages of Kinsey from over half a century ago, his institute's from twenty-five years ago, and the Durex and Definitive Penis surveys from the last year of the millennium (only three-tenths of an inch apart, top to bottom, after all) we arrive at 6.25 inches, with a circumference of just under 5 inches being pretty consistent in all surveys; and that surely seemed as definitive as you can get, except that in 2001 Lifestyle Condoms (on the same mission as Durex) carried out the only large-scale study *not* to rely on self-measurements – and turned the penis issue on its head. After getting three hundred volunteers to submit their aroused manhood to the attention of two tape-wielding nurses under the constant supervision of a doctor, Lifestyle reported the average erection to be 5.8 inches – about half an inch less than the above averaged averages. It's worth noting that five years earlier two small-scale studies (one in Germany, one in Brazil) had pharmacologically induced erections in volunteers and both had averaged out at 5.7 inches. Even more startlingly, the same year the *Journal of Urology* had published the findings of a study in which eighty normal men of various ethnicities had also been pharmacologically aroused (the object in this case was ultimately to help in counselling others considering penile augmentation) – and arrived at an average of 5.08, almost three-quarters of an inch less than Lifestyle's.

The medical profession continues to measure penises; between 2007 and 2010 at least fifteen different studies were published, all of them hands-on. What now seems to be the focus of attention is the likelihood that men who know or think they are below average are unlikely to volunteer to be sized up, or allow themselves to be, meaning that averages could be lower than those recorded (allusion to which hypothesis might, in some circumstances, stand a small-penised man in good stead). What is incontrovertible is that where men and their penises are concerned there are lies, damned lies, and self-measurements.

No one appears to have conducted any research to show whether gay men are more prone to exaggeration than straights but, certainly, size is an even greater issue in the gay community than in the community at large – gay men, having their partners' penises to think about as well as their own, take penises even more seriously than other men, which is saying something. The issue became very heated during the nineties after three researchers analysed twenty-five years of data from the Kinsey Institute and concluded in an edition of the *Archives of Sexual Behavior* that the average gay man's penis was longer than the average heterosexual's – at 6.4 inches, the equivalent of the average black man's. Much excitement was generated in the gay community, with black gays accused of being more 'sizeist' than their gay white brothers.

A final word on this matter of measurements. While most researchers run the tape along the upper surface of the erect member some, curiously, favour the under surface and a few what is called 'stretched penile length' – the stretched flaccid penis having been shown to more or less equal the length erect. The second method may be of benefit to a corpulent man with a substantial belly overhang, which curtails his ability to protrude (according to the Indiana University Medical Center in Indianapolis for every thirty-five pounds of weight gain the prepubic panniculus – the pad of fat – encroaches another inch on the penile shaft); the last may be of benefit to any man prepared to make his eyes water for a few extra millimetres against the ruler.

Reality and rumour

In Alan Bennett's play *Kafka's Dick*, the Jewish-Austrian writer Franz Kafka is spirited back to the present day with his father, Hermann, and is shocked to discover how famous he has become since his death. But so is Hermann, who is jealous and decides that the best way he too can achieve fame is to get Franz to tell the world how supportive of him he was in life – a rewriting of biographical truth. If Franz does not, Hermann threatens,

> I tell the world the one fact biographers never know.
> I reveal the one statistic every man knows about himself

but no book ever reveals. You see, it's as I say, we're
just a normal father and son, but what's normal?
My normal (*Indicates about 8 inches*). Your normal
(*Indicates about 3 inches*).

As far as his broad argument is concerned Hermann is either
lying or misinformed: writers like almost everyone are delighted
to allege that the penises of the famous are of such dimensions
as to attract admiration or ridicule. Virtually all evidence about
size is anecdotal, of course. Yet some claims come from so
many sources that they seem beyond dispute. The seventeenth-
century monarch Charles II was nicknamed 'Old Rowley' after
a prodigious sire he kept stabled at Newmarket – not just for
the number of his offspring (fourteen acknowledged bastards)
but because he was formidably hung. In 1663 the diarist Samuel
Pepys records Sir Thomas Carew telling him

> That the King doth mind nothing but pleasures and hates
> the very sight or thoughts of business; that my Lady
> Castlemaine rules him; who he says hath all the tricks of
> Aretine that are to be practised to give pleasure in which
> he is too able having a large —

The scatological poet John Wilmot, the 2nd Earl of Rochester,
was once banned from court because, when drunk, he mixed
up two poems in his pocket and gave the wrong one to the
king. 'A Satire on Charles II', which he did not intend to place
in the royal hand, contains these lines:

> Peace is his aim, his gentleness is such,
> And loves he love, for he loves fucking much.
> Nor are his high desires above his strength:
> His sceptre and his prick are of a length.

In the 1940s and '50s, the waiters of Paris referred to their restaurants' giant peppermills as 'Rubirosas', after the international diplomat-playboy Porfirio Rubirosa, whose conquests included Zsa Zsa Gabor, Ava Gardner, Jayne Mansfield, Marilyn Monroe, Kim Novak – and Eva Perón. According to one of his wives his penis was 11 inches with a 6-inch circumference, 'much like the last foot of a Louisville Slugger baseball bat' (*Too Rich: The Family Secrets of Doris Duke*). There's considerable anecdotal evidence to believe that Charlie Chaplin had good reason for calling his penis, as he did, the 'Eighth Wonder of the World' and to support the claim that at one time the three best-hung men in Hollywood were Frank Sinatra (his valet claimed his employer bought special custom-made underwear for concealment), Forrest Tucker and Milton Berle. 'What a shame – it's never the handsome ones,' Betty Grable lamented. 'The bigger they are, the homelier' – a slur on the likes of Valentino and Errol Flynn, widely regarded as probably the handsomest men ever to have appeared on the screen.

Today, in an age that has become so demanding of every detail of celebrity lives that no celebrity penis can go about its business in decent anonymity, there appear to be more 'bigs' than statistical distribution clearly specifies is possible – a situation more publicity mill than peppermill. But there's no disputing the size of the Valentino penis: he left hard evidence, a cast in solid silver, a memento to the actor Ramon Novarro of their affair. The Valentino penis swung both ways, as it were, principally away from the distaff, and was as long as his name. His full name, that is: Rodolfo Alfonso Raffaelo Pierre Filibert Guglielmi di Valentina d'Antonguolla. Another who left the measure of his manhood to posterity was Jimi Hendrix, captured by the Plastercasters, an arty cooperative of rock-star groupies that in the 1960s found something more intimate to collect than signed photos or autographs on their breasts. While these casts are originals (though Hendrix's has been lost), replicas of the penises of porn stars like Ron Jeremy and the late John Holmes, taken from their not-so-private parts, are on sale across America, and the world, via the Internet – for use as dildoes. Holmes, like Hendrix, was a 12, while at 9.75 Jeremy is a quarter-inch less than Valentino. 'They say 10 in some press and I say fine, I'll keep the extra quarter-inch,' Jeremy told *Onion* magazine. And by the way: his friend Holmes used to call him Little Dick.[4]

It's never mattered that rumours about penises being big or small are just as likely to be untrue as not for them to be relished as fact. Despite more than half a century of research that irrefutably indicates there is no justification, humankind continues to have a sneaking suspicion that short men are frequently over-endowed as some kind of compensation for a lack of height. Thus it continues to be written, for example, that the dwarfish painter Toulouse-Lautrec was known as the Coffee Pot in Paris brothels because of his huge member – though Julia Frey's life of the nineteenth-century painter carries a photograph of him in the nude captioned: 'Note short legs; genitalia and cranium appear normal.' The law of compensation, another common suspicion, goes into reverse for other small men whose deeds are the biggest thing about them – the 'bastardised conception ... great man, small member', wrote biographer Frank McLynn in defence of Napoleon Bonaparte.

The first suggestion that Napoleon's penis was 'abnormally small' came from assistant surgeon Walter Henry, one of the five doctors who conducted his autopsy. Conceivably at death it was: modern medical opinion is that Napoleon died of chronic arsenic toxicity from years of taking poisonous medicines, which would have atrophied his genitals just as it left his body hairless and made him fat to the extent of having breasts. But as McLynn emphasised, there was nothing in the emperor's earlier life to indicate any abnormality:

> As a man who liked to portray himself as a rough and ready soldier, Napoleon several times appeared in the nude in the presence of his troops . . . 'If O'Meara [his physician] writes a diary, it will be very interesting. If he gives the length of my cock, this would be even more interesting.' This hardly sounds like a man worried that posterity would laugh at him, and indeed O'Meara did produce a journal and made no use of Henry's 'astounding revelation'. Besides, even if we could imagine a substantially under-endowed man as a compulsive womaniser – which Napoleon was – his bedmates would surely have spoken of this interesting aspect of his anatomy.

There are times when people are prepared to ignore the evidence of their own eyes to ridicule a man's penis. It happened to the actor Jude Law in 2005 when he was caught cheating on Sienna Miller with the nanny, thereby setting himself up as a 'love-rat' and therefore deserving of a kicking. Shortly after, he was photographed changing out of swimming trunks on the veranda of his mother's house in France, and can't have been entirely surprised to have his perfectly respectable appendage described as a 'meagre manhood' and a 'puny package' – alliteration is the tool of derision.

And what of Franz Kafka? If Law was hard done by and Napoleon traduced, then think of poor, hypersensitive Kafka, a man with bad lungs, a hypochondriac array of other ailments, a fixation about masticating his food, an inability to form lasting attachments with women, and an intense love–hate relationship with his bullying father – but, as far as anybody knows, with a penis in no way out of the ordinary. Over 15,000 books have been written about Kafka and none save one has had anything to say on the subject that gives Alan Bennett's play its sustaining joke. The exception is a work by two psychologists at the University of North Carolina who analysed everything Kafka wrote and concluded on that basis alone that a small penis was at the root of his problems!

Given that they know they inevitably lay themselves open to ridicule about the size of their genitals, however normal they may be, we have to admire actors brave enough to appear naked on stage. When comedian Eddie Izzard appeared in the buff as Lenny Bruce in the West End, one newspaper quoted someone in the audience as allegedly saying 'He's obviously not being paid by the inch, is he?' Another comedian, Frank Skinner, found himself similarly ridiculed when he co-starred with a tortoise in the play *Cooking with Elvis* and was quoted as purportedly saying that 'being on stage with something small and wrinkly did not bother him. And playing alongside a tortoise was a nice change, too.' The classical actor Ian Holm had to suffer a critic's sneer when for the first time in his theatrical life he took off his clothes playing King Lear. But he had his revenge in his autobiography:

> Of my stage nakedness, there was little comment, apart from . . . Mark Lawson, who mentioned the shrivelled size of my manhood when I had to wade naked through a pool of cold water. Even disregarding Lawson's own physical shortcomings (the liver lips, the pudgy, plasticine face, the old man's prematurely balding dome), I am pretty sure his own equipment would also have dwindled after a cold bath in front of several thousand people. (*Acting My Life*)

While most mockery of men's penises comes from other men,

women are adept at verbal downsizing. Paula Jones, to whom Bill Clinton dropped his trousers, said nothing more scathing than that 'he wasn't very well endowed', but by the time her aggressive female lawyer was interviewed, the presidential penis had shrunk. Referring to claims of several affairs, she commented, 'If he did have sex with those other women they wouldn't have noticed' – mirroring Fanny Hill's remark about a client being 'of a size that slips in and out without being much minded'. It is the ultimate disparagement and even more withering when delivered by a woman with personal experience of the penis in question. A former mistress of a former British Tory minister, the rotund Lord Soames, delivered a dagger thrust to the groin with the comment that sex with him was like 'a cupboard falling on top of you with the key sticking out'. An even more devastating dismissal came from the ex-lover of then British deputy prime minister John Prescott whose manhood, she declared, was decidedly marginal – the size, in fact, of a chipolata. The *Sun* newspaper gleefully showed a photo of a two-inch cocktail sausage with the caption: 'Actual size'.

Vindictiveness may account for some allegations, but it's well to bear in mind that, as a rule, women see only one erection at a time, which denies comparisons ('How huge is huge when you have no frame of reference?' asks Isadora Wing, the heroine of Erica Jong's *Fear of Flying*), and in urgent circumstances, which militates against detailed linear appraisal; and that a study of sexual relations shows that, while women in love are apt to consider a lover's penis bigger than it is during a relationship, they consider it smaller after the relationship wilts and the parting is acrimonious. Disillusioned with her dull academic husband Graham, in Julian Barnes' *Before She Met Me*, his wife Ann looks at his genitalia as he sleeps naked on their bed, puzzling that so much trouble could be caused by 'so trifling' a thing: 'After a while, it didn't even look as if it had anything much to do with sex . . . it was just a peeled prawn and a walnut.'

HUMAN PRIMACY

Objectively, even big human penises are small, other than in comparison with other human penises; but virtually all human penises are big in comparison with those of the other 192 primate species. Flaccid, the penis of the gorilla and the orang-utan, both with much bigger bodies, is virtually invisible; erect, it reaches 1.5 inches or less; the chimpanzee, man's closest relative (sharing 98 per cent of his DNA) achieves an erection twice that of the other two apes but still only half the average human one. Why, comparatively, man's penis is so disproportionately large is a question that engages a clutch of disciplines including archaeology, anthropology and zoology, as well as the evolutionary, psychological and sociological branches of biology. Collectively they remain at a loss to provide what is known as 'ultimate causal explanation'.

The consensual 'ological' view is that when man's hominoid ancestors came down from the trees 4 million years ago, their penises were of a size with the apes – 'vanishingly small', according to archaeologist Timothy Taylor (*The Prehistory of Sex*). Then, however, when upright walking swivelled the sexual focus from rear to front of both sexes, a focus that was intensified by the loss of the majority of bodily hair other than in the genital area, the penis began the process of 'runaway selection'.

Feminists incline to the view that it happened because females wanted it that way; that when *femina* became *erecta*, the angle of the vagina swung forward and down, moving deeper into the body, obliging the penis, as Rosalind Miles put it in *The Women's History of the World*, to follow the same principle as the giraffe's neck: 'it grew in order to get to something it could not otherwise reach'. On the other hand, the big penis may have evolved because that's what possessors wanted – a greater attractant to potential mates and a more visible means of warning off rivals. A big penis also increased the male's chance of inseminating a female who was having sex with other males, by getting closer to the cervix. There are objections to such theories – not least that other primate males have continued to propagate their species with considerably less at their disposal. As to the theory that the penis grew to assist humankind's imaginative variety of sexual positions, orang-utans and chimpanzees, particularly the pygmy chimpanzee or bonobo (a separate species, found in the Congo, which has a more upright gait and a more 'human' skeleton), are equally imaginative in their coupling – and they can do it swinging from trees while man only talks about doing it swinging from chandeliers.

But if science cannot say definitively why man's penis is so big, it does have an explanation as to why his testicles are the size that they are.

In the early 1980s the evolutionary psychologist David Buss caused widespread excitement among the 'ologies' with the hypothesis (in *The Evolution of Desire*) that the more promiscuous a primate species, the larger the testicles of the males belonging to it – penis size, he surmised, was less relevant in achieving impregnation of a female having sex in rapid sequence with other males than being able to produce the most copious and frequent ejaculate. Subsequently, British scientists weighed the testes of thirty-three primate species, including man, to assess the testicle–promiscuity link. Interestingly, by this measure, the human male, the primate with the biggest penis, was not the king of the swingers: his testicles, together weighing 1.5 ounces, bore no comparison with those of the chimpanzee, which weighed an astounding 4 ounces, a three-times higher testes-to-body-weight ratio than humans. And the mighty gorilla, the primate with the smallest penis? Again he trailed the field, his testicles little more than half the weight of man's. As Buss pointed out, the gorilla, with his monogamous harem of three to six females, faces no 'sperm competition' from other males. On the other hand, the promiscuous common chimp has sex almost daily with different females and the even more promiscuous bonobo has sex several times a day.

Somewhere between gorilla and chimp comes man, neither entirely promiscuous nor entirely monogamous, his penis evolved far beyond those of his distant ancestors but his testicles or at least their firepower probably reduced – his sperm production per gram of tissue is considerably less than either chimps or gorillas, leading to the 'ological' view that, as expressed by Lynn Margulis and Dorion Sagan (*Mystery Dance: On the Evolution of Human Sexuality*), he once, when the business of insemination was a contest, had a bigger 'testicular engine'.

As with all body parts there are racial variations, a subject on which interest focused after Buss's theory of sperm competition became known and Jared Diamond described it as 'one of the triumphs of modern physical anthropology'. But measuring testicles was hardly as easy as measuring penises. A finger and thumb appraisal is wildly inaccurate: in the folds of the scrotal sac, testicles skitter out of grasp as easily as a bar of wet soap. Even measurement with an orchidometer (a specialised kind of callipers) is difficult – which is why scientists began to accumulate their data at autopsy. The findings confirmed what had been previously regarded as the case on less systematic analysis: that there is no demonstrable difference between the testicles of blacks and whites but that those of Asians are smaller. The extent of the difference, however, stunned the scientific community. It was more than twofold. As Diamond reported in a paper published in *Nature* magazine, where white and black testicles weighed an average 21 grams (there being 28 grams to the ounce), Asian testicles weighed 9 grams – the weight of the testicles of twelve-year-old white and black boys.

Men evidently equate testicles with manly courage (having balls or, as the Spanish-loving Hemingway preferred, 'cojones'), but, considering that the testicles are the manufacturing plant that helps achieve the Darwinian goal of procreation, they are surprisingly indifferent as to what size theirs are. Put that down, perhaps, to the fact that the testes are not truly visible, and that they have to play peek-a-boo with the penis in front of them, on which men lavish all their attention.

For the record the average black or white testicle is fractionally less than 2 inches long by 0.8 inches wide and is 1.2 inches in diameter, though some are half that and a very few up to half as big again, the largest having just over twice the volume of the smallest (Jane Ingersoll in Rick Moody's *Purple America* views Radcliffe's testicles as 'little cashews, not those asteroids some of her boyfriends have unveiled to her'). Taller and heavier (not obese) men tend to have big testicles, but this is a weak correlation – and there is no correlation to penis size. Hardly surprisingly, men with larger testes manufacture more sperm per day; and they ejaculate more frequently.

Testicular research of a more sociological kind has deduced that men with large testicles are likely to be more unfaithful, the converse being true of men with small testicles. A woman seeking a reliable long-term partner might be advised to invest in an orchidometer.

AESTHETICS, FUNCTION AND WOMEN

It's doubtful that any penis bears more than a passing resemblance to most of those on Grecian statues. Lifelike in the depiction of all other anatomical detail, the Ancient Greeks so idealised the penis (and indeed its attendant accoutrements) that they tidied up the imperfections. Flesh and blood penises are unlikely to be dainty, slim and pointy-tipped as are those in Grecian art or that of the Renaissance, which was enamoured of the Grecian tradition (Michelangelo's Adam on the Sistine Chapel ceiling, like God of a weightlifter's build, is barely sufficiently endowed to propagate the human race); and testicles are virtually never symmetrical and hang in the same horizontal plane (the perfectly matched brace of a king, as the *Brihat Samhita* has it) except, perhaps, when tightened by cold or fear. If one is frank about it, penises at rest generally appear unbalanced in one way or another, the scrotum in its normal state hangs pendulously like an avocado withered on the branch ('avocado' comes from the Aztec for scrotum), the testicles within it unequal, the right, with few exceptions, being larger and the left (because the spermatic cord is longer on that side) hanging lower, irrespective, curiously, of the 'sidedness' (right-handed, left-handed) of the possessor; according to the estimable male outfitters Gieves & Hawkes, some 80 per cent of men find it more comfortable to dress to the left.

Contemplating 'that capital part of man [and] that wondrous treasure-bag of nature's sweets', John Cleland's Fanny Hill concluded that they 'all together formed the most interesting moving picture in nature, and surely infinitely superior to those nudities furnished by the painters, statuaries, or any art, which are purchased at immense prices'; Lawrence has Connie Chatterley laud Mellors' genitalia as 'the primeval root of all full beauty.'

Sadly, this is transference of masculine wishful thinking. Some women may agree, of course, including the American artist Betty Dodson who once did sixteen drawings of male genitals 'so men could see all the wonderful variations in their sex organs' (Sex for One); however, as she describes the involved penises as 'Classical Cocks', 'Baroque Cocks' and 'Danish Modern Cocks with clean lines' it may be that her enthusiasm got the better of her. The prostitute in Tama Janowitz's Slaves of New York encounters all kinds of penises including some that are 'enchanted, dusted with pearls like the great minarets of the Taj Mahal' – which is almost as rhapsodic. At the opposite end of the spectrum, some women view male genitals with positive distaste, like the poet Sylvia Plath: 'old turkey neck and gizzards'; or, like Jane Ingersoll in Moody's Purple America: 'the ugliest anatomical part there is, next to goiters'.

Perhaps in expressing a view somewhere between extremes Esther Vilar (The Manipulated Man) speaks for most of her gender in saying that 'To a woman, the male penis and scrotum appear superfluous to the otherwise symmetrical male body' (considering the pandemic of obesity, 'symmetrical' being a theoretical concept, but let that pass). Certainly virtually all women find the female body, unencumbered by external sexual plumbing, infinitely more pleasing aesthetically; as Molly Bloom muses in her pre-slumber reverie, the female statues in the museum are 'so beautiful of course compared with what a man looks like with his two bags full and his other thing hanging down out of him or sticking up at you like a hatrack no wonder they hide it with a cabbageleaf' (*Ulysses,* James Joyce).

Men's feelings about all this are confused and contradictory. Possessors' affection for their penis is so great it's unlikely were they to be asked to name either their crucial external organ or their largest that they would reply, their skin; only propriety, perhaps, prevents many from displaying a sign in their car's rear window: I ♥ MY PENIS. Yet pride is underlain by varying degrees of anxiety. Eric Gill suffered none of this; he confided in his diary that he thought 'A man's penis and balls are very beautiful things.' Others may agree and, like Sebastian in *Romeo and Juliet*, consider themself 'a pretty piece of flesh' (flesh, of course, a biblical euphemism). But most men probably think the journalist A.A. Gill wasn't off the mark in describing male genitalia as 'the gristly cruet set' and wonder, despite their affection, whether theirs are inherently ridiculous to behold – classic Adlerian fear of mockery.

'Does your penis horrify women?' shouted an *FHM* magazine cover line, playing on this insecurity, a feature inside ('Are you ugly downstairs?') asking four women to assess their partner's penis against others when they were all thrust through holes in a screen. If hardly scientific, the exercise showed that the women easily identified their partner's (a small proof, tangentially, of penile individuality) and expressed affection for it – but mostly because it belonged to their partner, not because it was an attraction in itself. And while they recoiled somewhat from the three unfamiliar organs ('like a snake that's swallowed a football', 'too much skin flailing around', 'something in a butcher's window'), they found all of them rather funny, 'theirs' included. Women do; penises per se can be seen as something the Creator doodled in an idle moment. 'There's nothing so ridiculous as a naked man,' the very proper actress Jane Asher once remarked, a sentiment echoed by Debora from Derby when she appeared with her boyfriend in a television series on foreplay: 'The mere sight of Dave's penis', she said, 'has me in stitches.'

But Simone de Beauvoir (*The Second Sex*) was undoubtedly right when she observed that a penis-possessor, while regarding the idea of another man's erection as 'a comic parody . . . nonetheless views it in himself with a touch of vanity'. In truth, she understated the case because a penis-possessor's erection – 'man's most precious ornament' (Eric Gill again) – is his lion's mane and his peacock's tail, the source of his identity, the psychological and physical centre of his being, the very badge of his masculinity. To the penis-possessor his erection is a thing as wondrous as the metamorphosis of a caterpillar into a butterfly, even a recurrent miracle. His erection is

> a marvel of hydraulic engineering. In its enduring, reliable and repetitive efficiency it may be compared to the Gatun locks of the Panama Canal, which since 1914 have been raising ocean liners with swift and safe smoothness to 85 feet above the Atlantic and Pacific swell. The unstoppable power of the penile mechanism matches in ingenuity the channelling of mountain torrents, which since 1910 have whirred the turbines brilliantly to electrify the lamps of innumerable distant towns. The clever simplicity of penile erection, in applying fluid pressure to achieve motive power, recalls the mechanics of the hydraulic ram, or of the water-mills once scattered across the land . . .

Penis-possessors would not want this positively Rabelaisian

logorrhoea from John Gordon's *The Alarming History of Sex* to be ironical. And what they want from women on their erect penis's behalf, and which almost certainly they cannot articulate, is *awe*. Awe is what all males in the animal kingdom crave, Lorenz Konrad, Nobel prize-winning zoologist and father of ethnology, extrapolated from his study of tropical fish – the 'cichlid effect' notion of physiology. This, from D.H. Lawrence, complete with tumescent thicket of exclamation marks:

> 'Let me see you!'
>
> He dropped the shirt and stood still, looking towards her. The sun through the low window sent a beam that lit up his thighs and slim belly, and the erect phallos rising darkish and hot-looking from the little cloud of vivid gold-red hair. She was startled and afraid.
>
> 'How strange!' she said slowly. 'How strange he stands there! So big! And so dark and cocksure! Is he like that?'
>
> The man looked down the front of his slender white body, and laughed. Between the slim breasts the hair was dark, almost black. But at the root of the belly, where the phallos rose thick and arching, it was gold-red, vivid in a little cloud.
>
> 'So proud!' she murmured, uneasy. 'And so lordly! Now I know why men are so overbearing. But he's *lovely*, really. Like another being! A bit terrifying! But lovely really! And he comes to me!–' She caught her lower lip between her teeth, in fear and excitement. (*Lady Chatterley's Lover*)

Connie Chatterley's reaction is exactly as it should be, men are likely to think at some level of their being (and approve of the thicket of tumescent exclamation marks too). Sadly again, what we have here is the projection of more male wishful thinking.

The penis erect, according to Esther Vilar, 'appears so

grotesque to a woman the first time she hears about it that she can hardly believe it exists'. A first encounter is not likely to improve the situation for, as Inge and Sten Hegeler gently put it, 'an erect penis bears no resemblance to the kind that they have seen on statues in parks or on small boys paddling by the seashore'. Isadora Wing is remarkably unfazed by her first encounter with a 'phallos' (like Lawrence, Erica Jong favoured the Greek spelling); indeed she is intrigued by its 'most memorable abstract design of blue veins on its Kandinsky-purple underside' (well, she is an arts major). But most women are more likely to find echoes of their own experience in an article written by Lorraine Slater for *FHM* magazine:

> The first time I actually saw a real, live dick with my own eyes will be etched in my memory for ever. I was 15 and a few Pernods over the eight, squashed against a wall with my new guy, when all of a sudden he tried to force my hand down his keks. For some reason, he wanted me to fondle a smooth, rounded growth near navel-level. As I looked down I saw a glistening, angry-looking peeled plum thing glaring at me from above his belt-buckle. 'Jesus,' I remember thinking, horrified. 'That's his bell-end?' My mind whirled. How the hell did it get up there? Why don't they warn you about the colour? And the gloss finish?

As far as Maggie Paley (*The Book of the Penis*) is concerned, you can say that in spades: 'it was as ugly as a monster from outer space, and it seemed to have him in its power'. Kinsey (*Sexual Behavior in the Human Female*) found that a very small number of women are so repulsed by the aroused member that their erotic response is forever inhibited, an unhappy situation in which the advice given to a character in Alan Ayckbourn's *Bedroom Farce* might be apposite: 'My mother used to say, Delia, if S E X ever rears its ugly head, close your eyes before you see the rest of it.' The vast majority of women, of course, come to terms with the reality of masculine sexual mechanics: a rite of passage. Being practical by nature, they see an erection for what it is, the reflex of a body part that is fit for purpose: having sex – even if they are likely to be in agreement with Esther Vilar in thinking that 'It seems incredible . . . that a man cannot withdraw his penis after use and make it disappear like the aerial on a portable radio.' Yet, as Susan Bordo (*The Male Body*) observes, 'What other feature of the human body is as capable of making the welling of desire, the overtaking of the body by desire, so manifest to another?' It is a matter of constant fascination, and flattery, that they themselves are instrumental in conjuring the penis into life.

Unsurprisingly women are intrigued to know what having an erection feels like from penis-possessors' side of the sexual equation and penis-possessors find that almost impossible to explain. 'It seems light and heavy at the same time, like a piece of lead piping with wings on it,' suggested Henry Miller (*Tropic of Cancer*); 'On the borderline of substance and illusion,' offered John Updike (*Bech: A Book*). Most men would say that words are inadequate. At its greatest intensity a man may feel he is all erection and, perhaps, like Boswell, feel a 'godlike vigour' in its possession. In her night-time reverie, Molly Bloom ponders what it would be like to be a man 'just to try with that thing they have swelling upon you'. It isn't really a serious proposition – penises are for women to share, and give back. A few years ago, a publisher asked thirty women to contribute to a book entitled *Dick for a Day* and Germaine Greer in her response spoke for womankind in writing: 'The best bit would be getting rid of it.'

Is penis size important to women? This female contributor to *FHM* magazine was in no doubt:

In case you're one of those guys who's been mollycoddled by a sympathetic girlfriend, the question 'Does Size Matter?' is not up for debate. The jury delivered its verdict on that long ago, and yes – it bloody well does. Cocks don't do handstands, cook gourmet meals or speak Urdu. They go in-out, in-out. Size matters!

The corollary, of course, is that some women positively dislike the large penis, like Sandra Corleone in Mario Puzo's *The Godfather*:

> When I saw that pole of Sonny's for the first time and realised he was going to stick it into *me*, I yelled bloody murder . . . when I heard he was doing the job on other girls I went to church and lit a candle.

But what of the majority of women? Da Vinci thought that 'woman's desire is the opposite of man's. She wishes the size of the man's member to be as large as possible, while the man desires the opposite for the woman's genital parts.' Leaving the second observation aside, da Vinci was contradicted on the first by Masters and Johnson, who in the 1970s concluded that size was unimportant to a woman's sexual satisfaction. Researchers, now, are in some disagreement with that finding. If a penis is bulky (as opposed to lengthy), it makes greater contact with the outer parts of the vagina and therefore, it's thought, sends vibrations to the clitoris, the trigger of female sexual gratification. The ongoing debate is whether this is physiologically significant – or whether the bulky penis is only a psychological predilection. What is not up for reassessment, however, is that penis length is of secondary importance. 'If things aren't too bad in other ways I doubt if any woman cares very much,' the American playwright Lillian Hellman (*Pentimento*) observed.

As Alex Comfort makes clear in *The Joy of Sex*, 'female orgasm doesn't depend on getting deeply into the pelvis'. There's a good purely physiological reason: only the first two inches of the vagina are rich in nerve endings. The long and short of it, therefore, is that unless there's a startling abnormality, no erect penis is too small to surmount what the Chinese poetically call 'the jade terrace' and make adequate contact where it matters. And only in very, very exceptional cases is a penis too long, and virtually never too thick. A young man, weeks away from marriage, wrote to Kinsey explaining that his penis was 7.25 inches erect and 6 inches in circumference, which led him to be 'afraid that my organ is too big for intercourse with an average woman'. Kinsey wrote back, 'We have never seen a solitary instance in which the dimensions of the penis caused any difficulty in intercourse. Certainly we have records of successful adjustment where the penis measures two or three inches more than your own.' In being reassuring Kinsey was not being entirely accurate: elastic as the vagina is, a penis 'rare' in the Kinseyan sense can touch the neck of the cervix and the posterior fornix, causing pain; a man with such a penis may have to wear the equivalent of an outsize tap washer to reduce his intromittence. That said, the experienced Phoebe was entirely accurate in telling the inexperienced Fanny Hill that she 'had never heard of a mortal wound being given in those parts by that terrible weapon'. The vagina is highly accommodating – it can give passage to a baby at birth, after all – and closes upon whatever size penis is presented to it.

Women, of course, do have preferences about body parts, just like men. Some, for instance, have a taste for the circumcised penis, finding it neater, some for the uncircumcised penis, because the foreskin is another element and that it rolls back on erection adds intrigue. But in the final analysis whether a man is 'Roundhead' or 'Cavalier' is likely to be of scant significance

– as is size. Whether women have a partiality for 'bigs', whether in one dimension or the other or both, this isn't likely to top their wish list. What does, if a woman has feelings for the penis-possessor, is that she has no preferences – she accept his penis as it is, part and parcel of him: 'how innocently part of him it seemed, and not a harsh jutting second life parasitic upon him' in Updike's telling phrase (*Couples*). Anyway, a sensible man should give due cognisance to the fact that a woman is more likely to be turned on by broad shoulders and pert buttocks (should he have these) and by his ability to make her laugh than she is by his genitals. And, remembering Abraham Lincoln's dictum that a man's legs need only be long enough to reach the ground, he should be at ease with what he has. After all, what really matters is that he employs his penis to his partner's satisfaction as well as his own.

Yet research shows that almost without exception, irrespective of intellect, education and cultural or ethnic background, men ask women if the penises of previous incumbents of their bed were bigger. To appropriate a remark of the painter Ferdinand Delacroix (who was talking about canvasses), 'Men are always more given to admiring what is gigantic than what is reasonable.'

Whatever the evolutionary path of the human penis, Jared Diamond observes that it is four times bigger than biologically necessary, and 'as a structure [is] costly and detrimental to its owner'. And if the functionally unnecessary tissue 'was instead devoted to extra cerebral cortex, that brainy redesigned man would gain a big advantage'. The truth of that is theoretically undeniable – but one can guess that the men who would be prepared to make the swap would be few and far between. Indeed, as Phillip Hodson pondered in *Men: An Investigation Into the Emotional Male*, if the majority of penis-possessors had the option, they would only be satisfied 'with Beardsleyesque phalluses of such dimensions they need to be carried in both hands'.

PART ONE NOTES

1. In Shakespeare's *Antony and Cleopatra*, a soothsayer reads the fortune of two of Cleopatra's handmaidens and tells one, Iras, that her future is the same as the other, Charmian. Iras wheedles: 'Am I not an inch of fortune better than she?' Interjects Charmian: 'Well, if you were an inch of fortune better than I, where would you choose it?' To which Iras retorts: 'Not in my husband's nose.' A reader unfamiliar with the Elizabeth colloquialism is likely to assume that Iras, by innuendo, would like the addition exactly where she does not.

2. It intrigued the quintessential word manipulator, the Irish author James Joyce, that losing a word space gave the dubious synecdochic maxim 'The pen is mightier than the sword' an even more dubious, phallocentric meaning – not that he would have disagreed with Simone de Beauvoir's remark to all male sexual supremacists that 'the penis is neither plough-share not sword, only flesh'.

3. In the wake of the Durex and DP surveys and irritated that Asians were treated as one big group, an Asian website came into being with the ambitious hope of identifying variations between Japanese, Koreans, Chinese, Filipinos, Vietnamese, Cambodians, Laotians, Thais, Burmese, Malays, Indonesians, North and South Indians, Sinhalese, Pakistanis, Bengalis and Nepalese. It attracted only a few hundred replies and soon disappeared.

4. In the mid-1990s the honeymoon home video of American rock musician Tommy Lee and ex-*Baywatch* actress Pamela Anderson was stolen, turned up on scores of Internet sites – and Lee's penis for a time became the most viewed penis on the planet. Admittedly it was only of secondary or perhaps tertiary interest, but Anderson's assertion that it had her name tattooed on it, and 'when he gets excited it says, "I love Pamela very, very much, she's a wonderful wife and I enjoy her company to the tenth degree"', was clearly not entirely uxorious hyperbole, not that it stopped them subsequently getting divorced.

PART TWO

THE GIFT OF MASCULINE PERFECTION

You will always say of his membrum virile that
it is huge, wonderful, larger than any other; larger than your
father's when he used to get naked to take his bath.
And you will add, 'Come and fill me, O my wonder.'

Eighth-century Japanese pillow book

FROM BIT PLAYER TO LEAD

Towards the end of the sixteenth century, a fifteen-year-old French peasant girl named Marie was minding the family pigs when they escaped into a wheat field. Chasing them, Marie leapt over a ditch whereupon, according to the celebrated French surgeon Ambroise Paré, 'the genitals and male rod came to be developed'. In consternation Marie rushed to the physician and the bishop, neither of whom could offer help. Resigned, Marie renamed herself Germain and went to serve in the king's retinue. Years later the French essayist Michel Montaigne, on his way to Italy, stopped off to see the prodigy, who wasn't at home. He had not married, Montaigne was told, but he had 'a big, very thick beard'.

If Renaissance woman suffered anxiety that strenuous activity might make her prey to gender transformation ('there is still a song commonly in girls' mouths,' Montaigne noted, 'in which they warn one another not to stretch their legs too wide for fear of becoming males'), Renaissance man was made indignant, at the very least, by the tale of Marie/Germain and others like it. His conviction was that he was born to rule over woman and the Bible provided the evidence. 'The whole world was made for man,' opined the doctor and philosopher Sir Thomas Browne in *Religio Medici* (The Religion of a Physician) published in 1642. 'Man is the whole word, and the breath of God; woman the rib and crooked piece of man.' So did biology. In the first instance, as God had made man in his own image it followed, ipso facto, that God was a fellow penis-possessor, and woman, as a non-penis-possessor, was by definition lesser; in the second, medical authority held, as it had for more than a thousand years, that all foetuses were male: infants that emerged as female had simply failed to achieve masculine perfection. The reproductive organs of the female were male but in a defective state: the uterus was the scrotum, the ovaries the testicles, the vagina the penis and the labia the foreskin. These, save the last, had remained inside the female's body because she had generated insufficient heat to thrust them outside – a process seemingly not unlike turning a washing-up glove inside out.

To Renaissance man, the penis was God's ultimate gift; and that a woman might suddenly acquire one was not only an affront to God and to the natural order but to rightful penis-possessors. To Leonardo da Vinci the human body – the male, penis-possessing body – was even an analogy for the very workings of the universe, as depicted in his famous drawing of Vitruvian Man.

You can blame the Ancient Greek physician Galen for woman's inferior status in Western thought. It was he who erroneously theorised that there was a single model of human physiology, though he'd never seen inside the human body (he'd only dissected dogs and pigs). 'Turn outward the woman's [sexual organs], turn inward, so to speak, and fold double the man's, and you find them the same in both in every respect,' he wrote. And until the Enlightenment in the seventeenth and eighteenth centuries showed he was talking, well, cock (or balls, if you prefer), there simply were no words for female plumbing. Not surprisingly it did strike Galen that it was jolly useful that half the human race were botched males, allowing, as it did, childbirth, not to mention the pleasure of sexual intercourse. The Creator, he thought, would not 'purposely [have done it] unless there was to be some great advantage in such a mutilation'.

Was woman's inferiority ever thus? According to some interpretations of prehistory, the situation was very different: woman ruled over man, because she was deemed enchanted: mysteriously, monthly, she bled and yet healed herself; she produced new life from her own body. Be that as it may, once man realised he was necessary for conception to take place, non-penis-possessors became relegated. Until relatively recently, the consensual view was that this happened only when herdsmen and farmers began to control animal sexuality around ten thousand years ago and put two and two together about their own. Nowadays most anthropologists and evolutionary psychologists doubt the timeline, believing it insulting to our ancestors who, ten thousand years ago, had had anatomically modern brains for some one hundred and fifty thousand years.

All is conjecture. What is indisputable is that, if once the penis was only a jobbing extra in the story of life, by the time writing appeared around five thousand years ago, its name was already above the title of the film. 'Throughout all of history,' as Isadora Wing percipiently remarks, 'books were written with sperm, not menstrual blood.'

How true that is, is evidenced in mythology, those supernaturalised chronicles of the dawn of time and the birth of humankind. The Egyptian god Amun, for example, brought the world into being by swallowing his own semen and then spitting it out. The god Atum masturbated to create the Nile, while in Mesopotamia the god Enki, 'lifted his penis, ejaculated, and filled the Tigris', and had the stamina to move on to create the

Euphrates similarly (he dug irrigation ditches with his penis, too). Man was quick to create his gods in his own image: with a penis – three, in some manifestations of Osiris, another of the multiplicity of Egyptian penis deities. The Phoenicians even named their chief god Asshur, meaning penis, 'the happy one'. And man was already sizeist. The celestial rod of the Indian god Shiva shafted through the lower world and towered up to dwarf the heavens, so impressing the other gods that they fell down in worship.

There were, of course, goddesses as well as gods in the ancient world, and female generative power had its worshippers, but none was dominant in any major culture. The penis erect (ithyphallic, or straight up) ruled the roost and monuments fashioned in its likeness, usually in stone (the Japanese also favoured iron) sprang up like dragon's teeth across the globe. In Greece by the third century BC, the island of Delos boasted an avenue of massive erections, mounted on prodigious testicles, aimed skywards like cannon. Herms, the most famous and sophisticated phallic monuments – smooth stone square pillars that sported the bearded head of Hermes, the messenger of the gods, and, halfway down, an erect 'penis stick' – were placed at every crossroads to offer protection to the traveller, frequently anointed with wine and oil and constantly touched for luck. Herms stood at roadsides (as did the Japanese counterparts, the *dosijinas*), on every street corner and inside every home, as did their counterparts among the Egyptians, Hindus, Hebrews, Arabs and other Semitic peoples. The Scandinavians and the Celtic tribes of Europe placed phallic stones at strategic points on boundaries and at entrances.

The Romans adopted the herm and scattered it across their empire – and in echo had their gravestones carved with the likeness of their own head and genitals. Rome adopted the Grecian god Priapus, too (as Liber), a god permanently at awesome attention – though 'phallus', the word describing the penis in an erective state, derives from the god Phalles, whose cult of tireless sexual activity was rampant among young men. Greek, Roman and other cultures sculpted penises on the walls of their cities, houses and public baths to ward off bad luck, just as they protected their fields with replicas – and sometimes the real thing, removed from executed criminals and enemies. They decorated their household objects with phalluses, baked phallic cakes during festivals and wore phallic amulets – *fascina* in Greece, the word derived from yet another phallic deity – both to enhance their sexual potency and to protect them against harm.[*]

Images of phallic gods were carried in sacred processions, equipped (according to the Greek historian Herodotus, who visited Egypt in the fifth century BC) with movable members 'of disproportionate magnitude', to which were attached cords to control their movement. Many had a large eye painted on the glans, an early version of the all-seeing eye of providence, and women smothered them with garlands and kisses.

There is no exaggerating the reverence which once was bestowed on male genitalia, or the potency with which representations of them were considered to be imbued. Victorious Roman generals entered the city with a replica penis of great size suspended upon their chariot: symbol of victory but also a talisman against the envy of others. In the Middle East it wasn't unusual for a new king to eat the penis of his dead predecessor to absorb his power. In Kyoto, Japan, young men on the festival of a troublesome local goddess – who tried to break up young lovers – carried her image through the streets without wearing their loincloths, to keep her under control by the mere sight of their manhood. Greek and Roman men and women sometimes held seeds

---

* See Part Two Notes page 106

resembling testicles in their hand during sex to increase their lovemaking. Worldwide historical records and archaeological evidence show that in almost every culture young women were known to mount a stone or wooden phallus, the lingam in India, before their wedding night, giving their virginity to their gods (or sometimes, as the scornful Roman poet Lactantius wrote, that a god 'may appear to have been the first to receive the sacrifice of their modesty'); so did older women as a cure for infertility.

Among the Egyptians, Romans, Semitic Arabs and Hebrews, male genitals were so esteemed they were even a basis of civil law. Men clasped themselves and swore upon what they clasped. 'O Father of Virile Organs, bear witness to my oath,' the Arabs intoned. Romans made oaths in like manner, holding their testicles or 'little witnesses' – by extension not just to their virility but to their probity. The Hebrews went further, men making a pledge by clasping the genitals of the man to whom the pledge was being made. 'Put thy hand under my thigh,' Abraham in Genesis orders his servant, 'and swear by the Lord . . . that thou shall not take a wife unto my son of the daughters of the Canaanites.' There are other such instances elsewhere in the Old Testament; when, for example, Solomon was crowned king over all Israel, as related in Chronicles, 'all the princes, and the mighty men, and all the sons likewise of King David, gave the hand under Solomon'. Biblical translators, in a muck sweat over the issue, resorted to opaque circumlocution. There's no evidence that the Ancient Greeks testified genitally, but in Athens older men openly fondled the testicles of those not yet old enough to grow a beard when greeting them in the street. 'You meet my son just as he comes out of the gymnasium, all fresh from the baths, and you don't kiss him, you don't say a word to him, you don't even feel his balls!' complains a character in Aristophanes' comedy *The Birds*. 'And you're supposed to be a friend of ours!'[2]

Genital oath-taking appears not to have extended outside Roman and Middle Eastern cultures, but a related custom during the Middle Ages in Europe was for a woman who accused a man of rape to swear to the charge with her right hand upon the relic of a saint – and her left upon 'the peccant member'.

The über organ penis may no longer have been the arbiter of all things by the Renaissance but it was still fundamentally lord of all it surveyed. During the Enlightenment and beyond, however, advances in medicine and understanding of the human condition, as well as the lessening of superstitions, gradually eroded its power base. But at the beginning of the twentieth century came the entirely new field of psychoanalysis and the penis, if not exactly back on its pedestal, was again on the up, thanks to Sigmund Freud's conclusion that women suffer from penis envy. As small girls, he hypothesised, women see the penis of a brother or playmate, at once recognise it as the superior counterpart of their own 'small and inconspicuous organ', and fall victim to jealousy – an emotion only resolved when their subconscious desire for a penis 'changes into a wish for a man'. Freud also concluded that women are jealous of the penis as a urinary device. In childhood they witness small boys relieving themselves and flourishing their penis as if playing a game (the Jedi's light sabre has nothing on it) – and feel disappointment that they're denied the same pleasure of inventive manipulation (non-penis-possessors have to sit down, too; Freud felt sorry for them). Like others in his profession, Freud believed that this early experience made many women associate a garden hose with the penis because, as one of Havelock Ellis's patients explained, using a hose 'seems delightfully like holding a penis'.

No wonder, you might think, that women like watering the flower beds.[3]

SEMINAL INFLUENCES

The Athenians believed that the small penis was not only preferable aesthetically and sexually to anything bigger, but was also a superior delivery mechanism for human conception. Aristotle provided the scientific rationale. Having a lesser distance to travel along it, he argued, semen arrived in the hot condition required at its destination (another argument that still might stand a small-penised man in good stead, in some circumstances).

In the ancient world semen, like the penis itself, was regarded with something approaching wonder. Semen was the most precious of substances. The Greeks were convinced that the semen of an older man received by a younger in a homosexual encounter helped to build the recipient's manliness and passed on wisdom. The Romans celebrated a son's first ejaculation as part of the Liberalia festive holiday. Just as they protected their fields with replicas of penises, Romans, like the Greeks and other peoples, sprinkled semen to make the crops grow, a practice that still occurs in regions of Africa. As recently as the early part of the last century the New Mexico Zuni tribe were still leading one of their cross-gender priests on horseback onto the plains in the spring and masturbating him to ensure the return of the buffalo.

In East and West down the centuries, semen was considered to have magical properties. The Roman naturalist Pliny the Elder regarded it as a cure for scorpion stings; 1,400 years later, the Swiss-German physician and alchemist known as Paracelsus was convinced that a man could be created purely from semen, cutting out the middle man or, more accurately, woman. He wrote:

> Let the semen of a man be putrefied in a gourd glass. Seal it up in horse-dung for 40 days, or so many until it begins to be alive, move and stir . . . After this time it will be something like a man, yet transparent, and without a body. Now, after this, if it be every day nourished with man's blood, and for 40 weeks be kept in a constant, equal heat of horse-dung, it will become a true and living infant, having all the members of an infant born of woman.

When he was dying Paracelsus had his penis cut off and buried in manure, hoping to be resuscitated as a virile young man, a trick even less likely than turning base metal into gold.

Given the exalted view that men took of penises and semen, it's less than surprising that the ancient world was dismissive of women's part in the matter of conception. Both Hippocrates, the Greek father of medicine, and Galen believed that women, like men, produced semen but, while theirs played some part in the matter, it was cold, watery and insignificant; menstrual blood, on the other hand, was women's essential contribution, because it nourished the foetus. Its availability, Galen said, was due to women not being 'perfectly warm', as were penis-possessors, resulting in them having a surplus of blood left over from their comparatively cold bodily needs.

Semen in both Greek and Latin means seed and the word alone defined the respective part that the sexes were believed to play in making babies. As a Hindu text of around AD 100 decreed: 'The woman is considered in law as the field and the man is the grain.' Aristotle likened man to a carpenter and woman to the wood he worked. A woman, he also said, was 'a mere incubator'. 'The mother is not the true source of life,' Aeschylus has the god Apollo say in *Eumenides*. 'We call her the mother, but she is more the nurse: she is the furrow where the seed is thrust. He who mounts is the true parent; the woman but tends the growing plant.'

The question that exercised all ages was: where, within the male anatomy, did semen come from? The Sumerians thought it derived from the bones; the Egyptians, more specifically, the spinal column. Hippocrates taught that semen came from the brain direct to the penis, which Galen later amended, saying that it arrived from the brain to the left testicle, where it was purified and warmed to 'the peak of concoction', before being passed to the right to await usage. This led Aristotle, who believed that as spinal marrow and the brain were both white in substance, semen owed something to both, to conclude that a male child emanated from the perfect semen in the right testicle and a female child from the incompletely processed semen in the left. By extension, he concluded, boys grew on the right side of the woman's body, girls on the less-favoured left. The Japanese, Chinese and Hindus, meanwhile, also identified the brain as the seminal source, even believing that if a man refrained from ejaculating at the crisis moment he could reverse the flow of his vital essence and send it in the opposite direction to nourish its source (according to Indian tradition, a highly developed ascetic who sustains a cut bleeds not blood but semen). The Chinese and Hindus, like Galen, considered semen to be extracted in the testicles from blood – the Chinese said ten drops produced one drop of semen, the Hindus said forty.

No culture was of the opinion that the seed was manufactured in the testicles.

The view that all families in a manner of speaking were one-parent families did modify with the centuries, but the woman's input continued to be regarded as secondary. By the sixteenth century the conviction was that a man's semen transmitted not just life but a child's characteristics. Ideally, the male child possessed the man's complete identity. The view persisted right to the end of the eighteenth century that if a man fathered a weakling or a daughter, then the woman was likely to be at fault for not being submissive enough, or that the man's concentration had been broken and he'd been put off his stroke – as befell Tristram Shandy's father in the act of begetting his heir:

> *Pray, my dear*, quoth my mother, *have you not forgot to*
> *wind up the clock?– Good G–!* cried my father, making an
> exclamation, but taking care to moderate his voice at the
> same time, – *Did ever woman, since the creation of the world,*
> *interrupt a man with such a silly question?*
> (*The Life and Opinions of Tristram Shandy*, Laurence Sterne)

During the Renaissance and even later, medical thought in the main followed the Aristotelian tradition; one of da Vinci's anatomical drawings, based on dissection though it was, showed a seminal channel – which does not exist – from the spinal column to the genitals (the spinal connection meant that semen had long been called 'marrow'). But there were other assumptions. Nicholas Culpeper's book on anatomy in 1668 noted there were those who believed that semen was produced in the kidneys, 'because hot kidneys cause a propensity of fleshy lust'. Yet others believed that the ingredients of semen originated in various organs, combining at orgasm – this being concluded from the observation that orgasm appeared to involve the whole body.

Aristotle disagreed with Galen's two-seed theory; he believed all life came from eggs and taught that a foetus grew from the male semen that coagulated into an egg inside the female 'testes'. It wasn't until the late seventeenth century that the Dutch surgeon Regnier de Graaf discovered that an egg was indeed necessary to conception – but it was the woman's. Even though he realised that this egg travelled from ovary to womb he rejected the notion that female biology was directly responsible. What he was dealing with, he decided, was the *aura seminalis* – an ancient concept of philosophy that believed the 'nature, quality, character, and essence' of a future human being was not corporeal but spiritual, a kind of astral agent. What de Graaf decided was that the *aura seminalis* was corporeal after all – that it was 'the pungent vapour' of semen.

Three years later in 1678 another Dutchman, the microscopist Antonie van Leeuwenhoek, was the first to see the millions of spermatozoa in a sample of semen (which, he primly noted, came from 'the excess with which Nature provided me in my conjugal relations, not sinful contrivance'). He too rejected egg theory, proclaiming that a minuscule, fully formed human – a homunculus – resided in every single spermatozoon. So right did this appear to other men of science that soon there were further sightings of homunculi; and theologians pondered whether the contents of Adam's 'fecundating fluid' had had little humans inside little humans, like a set of Russian dolls.

Another two hundred years would pass before the basic story of human conception – ovum meets single spermatozoon, each contributes half the foetus's building blocks – was understood. When it was, it put paid to the ancient world's shibboleth that a woman's orgasm was essential for conception and that this depended entirely on the generation of heat. The Greeks believed that man, by virtue of his perfect warmth and the vigour of his intercourse ('the chaffing of the stones'), provided his heat naturally, making the 'permatic humour foam'; but woman, cold creature that she was, needed a man's ministrations to warm her up. Every culture continued in this belief (the Saxons even called the penis the 'kindling limb') and wasn't short of advice on how a man could achieve combustion. 'Handle her secret parts and dugs, that she may take fire and be inflamed in venery,' wrote John Sadler in 1636, 'for so at length the wombe will strive and wave fervent with desire.'[4]

The Chinese, believers in the heat/orgasm connection, were additionally keen on a woman's orgasm, for her own pleasure but more particularly for the man's well-being: her orgasm ensured that her yin reached maximum potency, thus strengthening his yang. 'The more women with whom a man has intercourse,' counselled *The Way, The Supreme Path Of Nature*, a philosophy that dominated Chinese thought for more than two thousand years, 'the greater will be the benefit he derives from the act.'[5]

RELIGIOUS PHALLUSY

At the height of British colonial rule in India, the wives of Victorian missionaries, merchants and military men were shocked to observe that every day a priest of Shiva emerged naked from the temple and went through the streets ringing a little bell – the signal for all the women to come out and kiss the holy genitals.

Westernisation has eroded the cult of the lingam, but India remains the only region of the world where penis worship, its rituals and its legendary narratives have continued from prehistory without interruption. In the more mystical tantric reaches of Hinduism and Buddhism, spreading eastwards through India's neighbours and the South Pacific, devotees are still said to regard themselves as merely 'phallus bearers', each the servant of his sex organ, which he regards as a separate living entity, a divinity, even, in its own right. Worshippers of Shiva seek not so much hydraulic assistance as an uplifting union with the world's seminal creativity.

As basis for faith, phallusism would undoubtedly strike

almost everyone today as either disturbing or ridiculous. But according to the respected orientialist Alain Daniélou, the first translator of the *Kama Sutra* since the Victorian Sir Richard Burton, 'there is probably no religion in which a substratum of the phallic cult does not exist'. And that includes Christianity. The wives of Victorian colonialists would have been reaching for the smelling salts had they been told that the cross, the very centre of Christian belief, is, in fact, a stylised representation of male genitalia, the upright the penis and the side pieces the testicles – a pagan symbol antedating Christianity by countless millennia. Penis and testicles are also the origin of the Christian Trinity, parodied by the novelist Henry Miller in *Black Spring*: 'Before me always the image of the body, our triune god of penis and testicles. On the right, God the Father, on the left and hanging a little lower, God the Son, and between them and above them, the Holy Ghost.'

The ancient world was convinced a man had to be 'complete in all his parts' to make it into the afterlife and took a dim view of a woman damaging the male compendium. Assyria even had legislation:

> If a woman has crushed a man's testicle in an affray, one of her fingers shall be cut off; and if although a physician has bound it up, the second testicle is affected and becomes inflamed, or if she has crushed the second testicle in the affray, both of her breasts shall be torn off.

Judaism took a similar view. In the only verses in the entire Old Testament that forbid a woman to help her husband, the Book of Deuteronomy warns that if two men engage in a street fight 'and the wife of the one draweth near for to deliver her husband out of the hand of him that smiteth him, and putteth forth her hand and taketh him by the secrets: Then though her hand shalt be cut off, thine eye shall not pity her.' Christianity didn't follow suit but followed Deuteronomy in balefully warning that 'He who is wounded in the stones or hath his privy member cut off, shall not enter the assembly of the Lord.'

So precious, indeed, were male genitalia that the Middle Ages believed a new pontiff was obliged to sit in a specially constructed chair, a *sedia stercoraria*, which had a circular hole through which a cardinal reached up to ensure His Holiness had the qualifications for the job, before solemnly announcing '*Testiculos habet et bene pendentes*' – He has testicles and they hang nicely.

The tale is simply too good not to be true, but sadly it is. That such a chair was once used in the papal enthroning is fact – but it was originally either a Roman 'dung chair' (commode) or birthing stool predating Christianity. The rest can only be called papal bull: a fiction stemming from another fiction, the existence of Pope Joan, an Englishwoman who, around 850, was supposed to have disguised herself as a man to enter holy orders and rose to the Church's highest office. It was as a result of this duplicity, the Middle Ages believed, that the chair test was introduced, to ensure no non-penis-possessor could try it on again – a case not of phallic worship but certainly an enshrinement of phallic supremacy.

Early Christians were slow to give up phallic worship; in fact, until the fifth century AD their phallic and monotheistic beliefs existed happily side by side: phalluses were carried in Christian religious processions and continued to be carved on churches; candle grease was dripped into the font at baptisms, representing semen. Over time the Church absorbed and transmuted aspects of phallicism and began trying to squeeze it out, but with scant success: at the beginning of the eighth century the theologian/historian the Venerable Bede wrote that Redwald, the most famous king in East Anglia, had two altars, 'one for Christ, one for devils'. The Church issued edicts galore against phallic practices and levied increasingly harsh penances, but failed to get the majority of the faithful to change their ways, or, indeed, many of the clergy. In the thirteenth century the minister of the church in the Scottish town of Inverkeithing was hauled before his bishop for leading a fertility dance round a phallic figure in the churchyard at Easter; in the fourteenth, the bishop of Coventry was accused before the pope of 'homage to the devil'.

Just after the end of the Second World War, Professor Geoffrey Webb, formerly Slade professor of fine art at Cambridge and at the time secretary of the Royal Commission of Monuments, was given the job of surveying those English churches that had been bomb damaged. In one, a blast had shifted the altar slab, revealing the interior for the first time in eight or nine hundred years and inside Webb found a carved penis. After investigating many other churches he found carved penises in 90 per cent of those dating up to around the Black Death that ravaged Europe in the mid fourteenth century.

Medieval Europe, devoutly Christian or not, simply continued to believe in the penis as a talisman and an insurance policy against bad luck. Priests directed parishioners with a problem in the sexual department (barren women, wives of impotent men, impotent men, men and women with venereal disease) to the local phallic stone, evidently thinking that touching it opened a channel of communication with a greater authority than they could muster. Across Europe and Britain people wore phallic amulets and women baked phallic cakes, just as in ancient Athens and Rome. And during spring planting and summer harvest, people took part in fertility festivals, as had happened throughout human existence, in Rome as the Saturnalia (which became notorious it got so out of hand). During such festivals, when sexual shenanigans were a cathartic release, and couples took to the wood 'to make green-backs' as Shakespeare put it, men went through the streets carrying wooden phalluses, prodding passing women with the tips or entering houses to prod the females (at one time the prodding was real, according to historical claims, and welcomed).[6]

And then there was the extraordinary ceremony known as the Feast of Fools, celebrated in December, an occasion that mocked the Church and frequently 'descended into lewdness and harlotry', with the clergy as well as some of their parishioners throwing off all their clothes. According to the French scholar and bibliophile Jean-Baptiste du Tillot, the bishops were powerless to stop these goings-on though they attempted to moderate them. Thus a ruling, in 1444, of the cathedral authorities in Sens in northern France, that 'those who wish to copulate go outside the church before doing so'.

The tussle between Christianity and phallicism continued. Pragmatically, the Church absorbed phallic cakes at Easter by ordaining they be marked with a cross (the phallic origins long forgotten) – lo, hot cross buns. Pagan feast days were transmuted into saints' days and the Feast of Fools became the Feast of the Circumcision. And Christianity fought penis with penis or, rather, prepuce, claiming to have discovered that of the baby Jesus, removed at circumcision – the only part of him, of course, that couldn't have ascended with him into heaven.

The earliest record of the holy foreskin was in AD 800 when the Emperor Charlemagne crowned Pope Leo III and presented it to him. Thereafter there was considerable rivalry for possession of the relic. Depending on what you read, there were eight, twelve, fourteen, even eighteen holy foreskins in various European towns in the Middle Ages. The most celebrated was sent in 1100 to Antwerp by King Baldwin I of Jerusalem, who purchased it during the First Crusade. Another, in the Abbey Church in Chartres, was borrowed by Henry V (of Agincourt) when his wife Catherine was about to give birth, to ease her labour. When asked in the twelfth century to rule as to which was the genuine article, Pope Innocent III declined, on the grounds that only God knew. However many holy foreskins were claimed, all but one were destroyed or lost during the Reformation and the French Revolution. The one that survived was carried in a reliquary through the streets of the Italian village of Calcata, north of Rome, until 1983 on the Feast of the Circumcision (though this was officially removed from the Church calendar in 1954). That year it was apparently stolen from the home of the parish priest. Popular opinion was that the report of the theft was the Vatican's way of ending the practice – which threat of excommunication issued in 1900 had failed to achieve.

Throughout the Middle Ages as throughout the centuries before them, the baby Jesus's penis was depicted by hundreds of artists – but only ever uncircumcised, as if the Son of God could not be envisaged anything but whole. The Church forbade the direct depiction of the adult Jesus's genitals (hence the unlikely loincloth while he hung otherwise naked on the cross) but during the Renaissance found it theologically acceptable when Dutch and German artists showed the suffering or crucified Christ with an erection. In the last Western age of true penis power, Christ's erection was a double image: of God's virility as the source of life and of the humanity of his Son as man – a double meaning, too, of the Passion (and, unintentionally, of Christ risen).

It took the Church until at least the eighteenth century to knock overt phallic practice out of the faithful. Up until then people in many areas of France, Belgium, Italy and Switzerland prayed to phallic deities. The Church's ingenious answer was to say that the statues of these deities (Foutin, Eutrope, Arnaud and Ters among others) were really those of Christian saints and even provided legends about them. And childless women were sanctioned to visit their local phallic 'saint'. This they did, not just to pray but to scrape his large wooden phallus, mixing the scrapings with water: a drink that would miraculously remedy their infertility – or put lead in their husband's pencil. When the phallus became too worn down, the priest renewed its dimensions with a few surreptitious taps with a mallet to the end behind the altar. In Montreux in Switzerland, a custom on the feast of the local phallic saint was for young men to mix their semen with water and try to get the girl they fancied to drink it.

That phallusism was alive and well – and doing good business – only two hundred or so years ago was verified in 1786 by the British envoy in Naples, Sir William Hamilton, who wrote to the president of the Royal Society explaining how in a little explored part of Isernia he found the peasants worshipping 'the great toe of Saint Cosmo'. And he deposited proofs of his findings with the British Museum.

During a three-day fair in September, Sir William had found, the relics of two phallic saints (the other, Damian) were carried in procession from the cathedral to an outlying old church, where 'a prodigious concourse of people' came carrying wax penises, 'some even of the length of a palm', purchased from street sellers. In the church vestibule, those who carried them – mostly, Sir William noted, women – kissed their votive offering before handing it, together with a piece of money, to a priest sitting at a little table. '*Santo Cosmo benedetto, così lo voglio,*' many murmured as they did so: Blessed Saint Cosmo, let it be like this – a prayer with several possible interpretations. At the church altar men and women uncovered any infirmity of their body, 'not even excepting that which is represented by the ex-voti', to be anointed by another priest with the 'oil of

Saint Cosmo'. The oil of Saint Cosmo was held in high repute, especially 'when the loins and parts adjacent are anointed with it'. On Cosmo's feast day the church got through 1,400 flasks of the stuff.

A clash of symbols

Are the spires, minarets and domes that rise above places of worship phallic symbols? Given that when the earliest of them were being erected religion hadn't shaken free of phallic worship, they almost certainly were, according to most authorities. Like the cross and the hot cross bun, and an almost endless list of religious artefacts claimed to have phallic origins (in some cases admittedly disputed), penile spires and minarets and testicular domes have long since lost their meaning. But to deny that that meaning was once very real would be, as the respected J.B. Hannay wrote (*Sex Symbolism in Religion*) in 1922, rather like discussing *Hamlet* without the prince.

Phallic symbolism is as old as phallic worship and almost everything with a resemblance to male genitalia in the natural or animal kingdom has been accorded phallic significance at some period in history – there's a lot of crossover with the metaphorical. The literature of every culture runs riot with phallic symbolism. In Greek mythology, Zeus' thunderbolt, Poseidon's trident and the caduceus of Hermes, not to mention the 'massy clubs' carried by the likes of Hercules and Theseus, the ancient world's superheroes, were all symbols of the potency and power of the penis, just like Norse hammers, Tibetan *dorjes*, Chaldean swords, Chinese dragons, the witch doctor's or wizard's wand and the monarch's sceptre (this reinforced by the 'witnessing' orb, topped by a cross for further reinforcement).

Many phallic symbols have been only of their time. The setting sun was in prehistory seen as the engorged tip of the penis plunging into the female earth and the rain that moistened and fertilised the female earth a kind of heavenly semen – something that appears in the oldest layers of many literatures. When moonbeams were considered phallic, women would not sleep in their light in case they were made pregnant by them. Before household door locks became commonplace, well-to-do ladies carried a 'chatelaine' key chained to their girdle – symbol of the authority of the household penis-possessor by proxy – to lift door latches, when a finger would have done the job as easily.

Time has neutered the majority of phallic symbols including the village maypole. A part of pagan fertility worship in pre-history and still a copulatory symbol in the Middle Ages, the maypole was burnt by evangelical Protestants at the Reformation, banned under Cromwell ('a heathenish vanity, generally abused to superstition and wickedness'), and while restored at the Restoration, lost any remaining sexual significance so that by the nineteenth century, which added the ribbons for dancing couples to intertwine, it was erroneously taken to be an innocent reminder of a Merrie England that never was.

By the nineteenth century phallic symbolism had largely dropped out of general consciousness. But psychoanalysis brought it back in a big way – by locating it as hardwired in the subconscious. In his *Interpretation of Dreams* Freud listed many phallic symbols common to history and added many others including neckties ('which hang down and are not worn by women') and 'balloons, flying machines and most recently Zeppelin airships' because they all shared the 'remarkable characteristic of the male organ . . . to rise up in defiance of the laws of gravity'.

In the rise of rock music during the 1960s, the anthropologist Desmond Morris identified the electric guitar as a new phallic symbol. The traditional acoustic guitar, he noted in *The Human*

Zoo, with its curvaceous, waisted form, was essentially feminine; the electric guitar, however, had affected a sex change:

> the body (now its symbolic testicles) has become smaller, less waisted and more brightly coloured, making it possible for the neck (its new symbolic penis) to become longer. The players themselves have helped by wearing the guitars lower and lower until they are now centred on the genital region.

And, of course, where the acoustic guitar is usually caressed at chest height, the electric guitar, manipulated at an erective angle, is repeatedly stroked violently in a manner that can be described as masturbatory.

What is phallic is sometimes only in the eye of the beholder. Habitually Salvador Dalí mentally superimposed three church belfries that were meaningful in his life to help him masturbate; when Aubrey Beardsley had a tooth pulled he made a sketch of it and wrote in his diary, 'even my teeth are a little phallic' (entirely coincidentally the American poet Walt Whitman described the penis as a 'tooth-prong'). Innocent objects often took on a phallic appearance to the Earl of Rochester when he was drunk. Weaving across Whitehall Garden after a night-time drinking session with the king and others, he saw His Majesty's most highly prized possession, the rarest sundial in Europe, made of glass spheres, screamed 'Does thou stand here to fuck time?' and smashed it to smithereens with his sword.

Modern psychiatry has for the most part turned away from phallic symbolism, deeming it subjective and unscientific, and so it may be. But for most people, at a superficial level, 'phallic' is a reflex descriptor for anything rigid and upright, be it flagpole or lamppost, tower block or skyscraper; the heroine of Amanda Craig's novel *Foreign Bodies* says dubiously on seeing an erection for the first time: 'It is, I suppose, the basis of a great deal of architecture.' And phallic symbolism remains beloved by writers of high-end literature, and of course filmmakers: tumescence (trains enter tunnels, rockets shoot into space, fireworks climb into the sky, wave-peaks race towards land), ejaculation (volcanoes erupt, champagne corks pop, fireworks spray, waves pound rocks or shore), and detumescence (hot-air balloons deflate, detonated chimney stacks topple, fireworks fall, waves withdraw). But archaeologists and anthropologists are conceivably the most committed of phallic symbolists. As anthropologist Richard Rudgley admitted a few years ago in a television programme looking for secrets of the Stone Age in the ruins of a Maltese temple built three and a half thousand years ago: 'It's an occupational hazard. We tend to see willies pretty much everywhere.'

Without doubt the best-known phallic symbol of modern times is the cigar. Indeed Freud, a cigar smoker who despite mouth cancer refused to give them up (yes, he affably agreed with his friends, smoking them was akin to homosexual fellatio), became so tired of hearing 'phallic' and 'cigar' in conjunction that he sighed: 'Sometimes a cigar is just a cigar.' But then, as the Bill Clinton and Monica Lewinsky affair goes to show, sometimes it isn't.

FLAUNT IT!

Given the Ancient Greeks' admiration for the penis, it's unsurprising that men exercised naked in the gymnasium (*gymnos* means naked) and took part naked in athletic contests. But in the fifth century BC, the period of the greatest flowering of Attic culture, foreigners were astonished to see that Athenian men, when young, habitually displayed their genitals in everyday life. While older men wore a tunic (*chiton*) under winter and summer cloaks, young men did not. And the light summer cloak (*chlamys*), which came only to the thigh, was frequently lifted by normal activity (not to mention a breeze). The Athenians did consider the head of the penis to be unseemly in public, which is why young men pulled their foreskin over it and tied it with a thong or held it closed with a circular clip called a *fibula*.

Small boys everywhere can be observed regularly delighting in pulling off their clothing to display their bud of flesh; and why not? asked the sexologist Alex Comfort, 'after all [penises] are some of the best things we've got'. If mature penis-possessors have an innate desire to do as little boys do, social convention ensures they do not, as it ensures that they refrain from the overt tactility that was also a constant feature of their earliest years (professional footballers the exception to this rule). When drunk, however, some penis-possessors have a compulsion that is irresistible. The list is long and grows. In 1581 John Harris of Layer Breton, Essex, was taken to court because he 'behaved himself very disorderly by putting forth his privities'; in 1590 Henry Abbot of Earls Colne, Essex, also appeared before the magistrates for undoing his breeches 'in his drunkenness claiming that his privities or prick was longer by 4 inches than one Clerke there'. In the following century Pepys recorded the trial of Sir Charles Dydley for debauchery after he had appeared drunk and naked in daylight on the balcony of a brothel,

> Acting all postures of lust and buggery that could be
> imagined . . . saying that here he hath to sell such a powder
> as should make all the cunts in town run after him . . . And
> that being done he took a glass of wine, washed his prick
> in it and then drank it off: and then took another and drank
> the King's health.

In our own time, the constantly inebriated actor Oliver Reed, who displayed his 'wand of lust' (which in moments of sobriety he admitted was nothing out of the ordinary) in bars, on planes, at parties and on television and film sets, once did so to a woman reporter who was interviewing him. In reply to her scornful: 'Is that it?' he said, 'Madam, if I'd pulled it out in its entirety I'd have knocked your hat off.' On another occasion when he exposed himself in a Caribbean bar the locals took the tattoo of an eagle's claws on his penis for a voodoo sign and he was forced to flee.

Men with a penis of considerable size don't need their

inhibitions loosened by alcohol to make the fact known or to demonstrate the evidence at the drop of a trouser. James Boswell, the to-be biographer of Samuel Johnson, in London from Edinburgh for the first time in 1762 and 'really unhappy for want of women', picked up a girl in the Strand and took her into a dark courtyard with the intention of enjoying her 'in armour' (he feared the pox). But neither he nor the girl had a sheath so they only toyed with each other and in his journal Boswell recorded, 'She wondered at my size, and said if I ever took a girl's maidenhead, I would make her squeak.' Prince Grigori Alexsandrovich Potemkin, the outstanding eighteenth-century Russian statesman and lover of the Empress Catherine, used to stride through the Winter Palace naked beneath his unbuttoned Turkish dressing gown, demonstrating that his reputation was not exaggerated; the priapic Russian holy man Grigori Rasputin, once accused in a packed Moscow restaurant of not being who he claimed to be, said 'I will prove who I am' and did – another whose reputation went before him.[7]

Eric Gill, the twentieth-century artist/sculptor and another journal keeper, like Boswell recorded a prostitute's comment about his size – in her case because 'it was too big and hurt her'. Gill habitually wore a short stonemason's smock without underwear and was wont, when showing visitors around his commune, to urinate in the grounds, which gave him the opportunity to display 'the water tap that could turn into a pillar of fire'. There is an entry in his 1925 diary concerning his secretary Elizabeth Bill: 'Talked to Eliz. B. re size and shape of penis. She measured mine with a footrule – down and up.' So proud was Gill of his endowment that he drew it constantly, made a carving of it, and used its proportions in wood engravings of Jesus who, he said, as man 'had to have proper genitals'. Proper genitals were what Gill gave the stone figure of the Shakespearean sprite Ariel when he carved him (accompanying the magician Prospero) above the entrance of the BBC's Broadcasting House. The BBC governors, gathered to see the work unveiled, were startled when Gill removed the tarpaulin behind which he worked – and ordered the sculptor back up his ladder to cut things back.

According to the autobiography of Hollywood mermaid Esther Williams, her one-time lover Johnny 'Tarzan' Weissmuller was so childishly delighted to be well hung that he lost no opportunity to flash his genitalia on and off the set, as did swashbuckler Errol Flynn, whose penis was such a heavyweight that his party trick was to play the piano with it. Yet another 'exhibitionist extraordinaire', according to his biographer James H. Jones, was Alfred Kinsey who

> seldom passed up an opportunity to show off his genitals and demonstrate his masturbatory techniques to staff members. One insider…told an interviewer that Kinsey 'had very large genitalia, and that means both penis and balls'. The man added, 'Several of the staff members used to say, "Maybe that's why he whips the goddam thing out all the time to show you the urethra or the corona".'

At least Kinsey had a quasi-scientific excuse; the American president Lyndon Johnson had no excuse at all, other than pride in his size. He loved to conduct official business while he was in the shower (his White House staff squeezed into the bathroom) and often emerged fingering the considerable presidential appendage saying, 'Wonder who we'll fuck tonight? . . . I gotta give Ol' Jumbo here some exercise.' (*The Years of Lyndon Johnson*, Robert A. Caro). Once, frustrated at a press conference at which he was being pressed as to why America was still fighting in Vietnam, Johnson pulled out his pecker (one of his favourite words), saying, 'This is why,' presumably because he thought it did his talking more coherently than he did. (If men do some of their thinking with their penis, it goes without saying that their penis talks to them, as Gray Jolliffe's cartoon character Wicked Willie – whose exploits have sold over five million books – so cleverly shows; the additional biological impossibility is neither here nor there.)

The aforementioned Scottish actor Ewan McGregor was only marginally less inarticulate than the presidential penis when he was interviewed after playing a rocker in the film *Velvet Goldmine*. For the role he had been required to moon at the audience; for good measure during shooting he flaunted not just his posterior parts but his anterior too. Asked why he had an irresistible urge to exhibit himself McGregor said:

> I don't go round thinking: Hey, I've got a huge cock, go
> on, show me yours and we'll compare sizes. But at the
> same time, when people ask me if I'd be keen to flash my
> willy if it was small, I always think: Well, how the fuck am I
> supposed to know?

Shortly after, he appeared on the cover of *Vanity Fair*, 'kilt akimbo', clutching a rooster – the quintessential cock o' the north.[8]

Accessorise – or aggrandise?

In areas of South America, Africa and across the Oceanic world, men who are otherwise naked wear penis sheaths (phallocarps, in anthropologist-speak). In some societies these are just caps over the penis head, but in others they not only cover the shaft but can extend as much as two feet, held erect by a thong around the waist or even the chest. Usually penis sheaths are made from bamboo or specially grown gourds (though in recent years old toothpaste tubes and Coca-Cola cans have extended the range) and are often bright red or yellow. Some societies have everyday sheaths, sheaths for festivals and sheaths for war – and sheaths for visitors. Most men possess a variety, varying in size, decoration (boar tusks, animal teeth or claws, feathers) and angle of erection, and wear them according to their mood. Penis sheaths reflect a man's status, warn off his enemies, enhance his penis's magic powers and, with luck, attract women to him.[9]

These were powerful reasons why, in the 1970s when, in the interests of public modesty, the Indonesian government attempted to ban men in the highlands of Papua New Guinea from wearing their sheaths, there was fierce and ultimately successful opposition. There was another reason why the men of the Dani tribe were deeply angered: they thought their sheaths *were* modest – which caused Marie, the wife of the anthropologist Jared Diamond, to describe them on her first encounter in the 1990s as 'The most immodest display of modesty I've ever seen.'

The psychologist John Carl Flugel at University College London saw all human clothing as a tug of war between 'the irreconcilable emotions of modesty and the desire for attention'. How the dictum applies to the codpiece, once the Western world's equivalent of the penis sheath, is open to debate. In any event, for more than a hundred years, as the Middle Ages merged with the Renaissance, the codpiece was the flamboyant focus of male attire and modesty never came into it.

Actually, the codpiece was rather more modest than what went before it – which wasn't exactly a fashion in itself, at least at first, but the consequence of one.

The mid fourteenth century was an age when clothing changed radically, the newly popular button allowing that of both sexes to fit more tightly to the body. The waistline of men's doublets dropped to the hips to make the body look longer and the hemline rose up the thighs to make the legs look longer. And herein lay the problem: men's woollen hose consisted of two separate legs, exactly like a pair of long stockings, but attached to a waist belt or, in the newer fashion, to lacing holes on the doublet's hem, with a space between the legs at the crotch. While the doublet had flared to the knees, men's braies (the medieval equivalent of underpants) had been bulky; now, they were much reduced so as not to spoil the closer-fitting shape of the shorter garment. And when a man sat down or mounted a horse, his braies poked through the opening, the necessary slit in them from time to time affording a glimpse – or more – of what was within.

Most men undoubtedly took care to mind the gap and maintained their decorum, but the common-sense solution was to sew the crotch shut. Yet when this happened, as paintings of the period indicate, young bucks dispensed with their braies altogether, their hose delineating and thrusting out their genitals in a forthright manner. Chaucer put it like this in 'The Parson's Tale' (related here in modern English):

> Alas! Some of them show the very boss of the penis and the horrible pushed-out testicles that look like the malady of hernia in the wrapping of their hose, and the buttocks of such persons look like the hinder parts of a she-ape in the full of the moon. And moreover, the hateful proud members that they show by the fantastic fashion of making one leg of their hose white and the other red, make it seem that half of their privy members are flayed. And if it be that they divide their hose in other colours, as white and black, or white and blue, or black and red, and so forth, then it seems, by the variation of colour, that the half of their privy members are corrupted by the fire of Saint Anthony, or by cancer, or by other such misfortune.

As stretched wool has a degree of see-through, which

increases markedly as it wears and the weave loosens, some penises, one supposes, looked like bank robbers in stocking masks. Be that as it may, young men were so pleased with themselves that they now went about in their shirtsleeves, discarding the doublet and wearing their money pouch at the front rather than at the side with their dagger swinging suggestively behind it – a pseudo-penis directing the viewer's gaze to the genuine article.

Men already had pseudo-penises on their feet, of course. You might think that the erroneous correlation between foot size and penis size is a modern one, but it isn't: it was a popular belief at least as far back as medieval times. And when shoes called poulaines (they originated in Poland), which had toes that tapered to a point of about six inches, became a craze across Europe, they presented an opportunity that men could not resist: they packed their toes with moss or wool, increasing their length to as much as 30 inches, kept them in an upright position with silken ties or silver chains fastened to their knees, and stood on street corners waggling them at passing women. The raciest fellows painted their poulaines flesh-coloured, just in case someone didn't get the connection.

The Church thundered about sin and proclaimed the Black Death and subsequent waves of bubonic plague (which wiped out two-thirds of Europe's population) were heaven's retribution for the obscenity of men's footwear and clothing. In 1482 the Commons petitioned Edward IV that

> No knight, under the estate of a Lord . . . nor any other
> person, use or wear . . . any Gowne, Jaket, or Cloke, but
> it be of such a length as it, he being upright, shall cover his
> privy member and buttokkes.

In a feudal world, lords were left to do as they pleased, as they were, more or less, where poulaines were concerned – Edward did limit the 'beaks' (points) of their footwear to no more than 24 inches, whereas those of lower standing were restricted to 12 and the masses to 6, exceeding which limits was 'upon pain of cursing by the clergy, and by Parliament to pay twenty shillings for every pair'. Such harsh sumptuary law (under which the penalty would have been beyond the majority's ability to pay) was intended to stop the fashion dead in its tracks and it did, though common sense must also have prevailed – poulaines might have been a gift to the genitally focused poseur but were almost impossible to walk in. Before the end of the century, they were gone

Edward had attached no penalty to his decree regarding men's hose, but the sewed-up crotch would never have lasted anyway: men couldn't put up with dropping their hose – first disentangling it from the waist belt under their doublet or unlacing the ties around it – every time they needed to urinate. Again common sense dictated the next fashion move: tailors inserted a simple triangle of material between the legs either stitched to the hose at the bottom angle with ties at the top two, or with ties at all three. And thus the codpiece (codd being Middle English for bag or pod) was created.

At this stage the codpiece was almost self-effacing, but not for long: men started bulking out their flap of fabric. For some, in all likelihood, it started as a joke; for those of a sizeist mentality, it became competitive. The flap evolved into a bulbous pouch, which then became increasingly padded, and more and more ornate, adorned with precious stones, strings of pearls, even little bells. Leonardo da Vinci was delighted that the penis was being given the regard he felt it was due, writing: 'Man is wrong to be ashamed of mentioning and displaying it, always covering and hiding it. He should, on the contrary, decorate and display it with the proper gravity, as if it were an envoy.' It was around the middle of the sixteenth century, some thirty years after his death, that the codpiece reached its height, blatantly shaped like an enormous, permanent erection – a boon, Rabelais wrote, to the 'many young gentlemen' whose 'fraudulent codpieces . . . contain nothing but wind, to the great disappointment of the female sex'. After that, the only way for the codpiece was down. During the Elizabethan age male fashion again changed dramatically, the trunk of the hose turning into something like puffy shorts, out of which, in diminishing size, the codpiece peered, before giving a valedictory wave.

It wasn't until the end of the eighteenth century and the beginning of the nineteenth – the period covering the French Revolution and the English Regency – that the penis again made a fashion statement across Europe, and the Americas. Knee-length breeches, which had long replaced hose, had introduced the 'fall front' – rather like a little apron that dropped down from the waist – for essential access. Less material was therefore required directly at the groin, and breeches, customarily in light buckskin, became so tight that for the first time in history tailors asked their customers which side they dressed, to allow some wriggle room to one side or the other.

Ballet had recently dispensed with loose-fitting garments for both male and female so that the audience could better appreciate their artistry and athleticism. Audiences at first had been taken aback by the frank forward thrust of the males' white tights, but had come to accept the sexual neutrality in the theatrical context (unlike the BBC governor who, in the 1930s, when ballet was first televised, suggested to the director-general that male dancers should wear two pairs of tights). Carnally minded Regency fops, however, inspired by the look, adopted flesh-coloured tights for everyday wear. A German visitor to London, quoted by Ivan Bloch in *Sexual Life in England*, saw that 'from a distance I really thought some inmates of Bedlam had escaped from their keepers and had put on only shoes and coats, leaving the rest of their bodies exposed'. In the same period, the costume worn by Spanish matadors was also being pared down to the classical 'suit of lights', the knee-breeches, like the tights of the dancer, sculpted to the body to emphasise the wearer's graceful virility – and the fact that, in the contest with the bull, the matador leads with more than his chin.

With the arrival of trousers as we know them in the mid 1800s, genital accentuation became less blatant but remained popular among younger men. Not without the ambivalence that characterised their age, which managed to be both libidinous and sexually repressed at the same time, many Victorian men chose to wear trousers more than necessarily delineating but then threaded a metal 'dressing' ring – the erroneously named 'Prince Albert' (see Note 9 page 108) – enabling them to anchor their penis with a ribbon or thin chain to their inside seam, thus lessening the visual impact. The Victorians passed on the taste for genital fashion flamboyance (though not for cock rings) to Edwardian dandies, after which it disappeared until the 1950s when working-class youths caricatured Edwardian styles, their trousers so tight they were obliged to walk with their penis in the upright position. 'Held firmly in its vertical posture by the snugly clinging material,' the anthropologist Desmond Morris wrote (*Intimate Behaviour*), 'it presents a mild but distinctly visible genital bulge to the interested female eye. In this way the young male's costume once again permits him to display a pseudo-erection . . . '

Trousers worn by (principally) young men during the 1970s were again cut high into the groin, leaving little uncertainty on the right or left issue, before styling for comfort and discretion became the norm, as it remains in the main, save for those addicted to the cowboy look, principally southern Americans of all sexual inclinations, and the gay community everywhere. Male anatomy being what it is, however, any man wearing trousers of light material and careless of the fact that when he sits the slack can become trapped beneath his buttocks may demonstrate a braggadocio he doesn't intend – as did the broadcaster Terry Wogan while appearing in his moleskin strides on television. In the circumstances it was doubly unfortunate that he was presenting *Points of View*, a title on which the newspapers gleefully punned in alluding to what they called Wogan's 'wardrobe malfunction'.

In what manner and to what degree clothed male genitalia have been exhibited, the intention generally has been to keep things in check. But an astonishing exception came in the 1980s when athletes took to wearing Spandex, known more familiarly as Lycra. The exceedingly thin and figure-hugging material was aerodynamic, developed to help athletes improve their sporting performance. But the constraining properties were negligible: penises bobbed and weaved with merry abandon in the track events, much as they might had their possessors been naked. And the considerable self of the British Olympic-gold sprinter Linford Christie, viewed in countless television slo-mo replays, became known to the nation as 'the lunchbox'. Several manufacturers of actual lunchboxes approached Christie to endorse their products, much to his annoyance. At the end of the 1990s when his running days were over, Christie was accused of having taken performance-enhancing drugs and instigated a libel action. During the hearing in the High Court the lunchbox was mentioned (irrelevantly to the case) and, possibly thinking it had something to do with cucumber sandwiches and unaware of what an entire nation knew, a bemused Mr Justice Popplewell sought clarification. Christie explained, adding that he was so fed up with people asking him how big his penis was he replied 'This big', stretching his arms to their full extent. If the learned judge wondered whether the gesture was meant only as a sarcastic response, he forbore from asking.

Christie's exceptional case aside, man seems to have a desire to draw attention to his clothed genitals, often by insinuating that less is more. It may not be a fixation, but it is inherent. Flesh-coloured tights weren't enough for some eighteenth-century fops: they added padding (they wore false calves too); twentieth-century pop singers likewise made a bolus of their groins, pressing a bunched handkerchief or rolled-up pair of socks into service, though the duo Wham! ingeniously used shuttlecocks. Some male catwalk models today have admitted to similar stratagems, which are no doubt employed by an unknowable number of men about their daily lives – who haven't; made-for-purpose 'package enhancers' are sold by many sex outlets. In 2010 Marks & Spencer, that bastion of middle-class fashion conservatism, began to stock 'frontal enhancement' underpants, an 'integral shelf' giving a claimed '38% increase in outline'.

From time to time, fashion historians have suggested that the codpiece, in its comparatively honest fraudulence, is due for a comeback. The idea has crossed the mind of several anthropologists including Desmond Morris and is never far from the mind of filmmakers, who rarely envision a future world without it.

In fact, the codpiece has been back with us, in a minor way, for the last sixty years – a staple of the worldwide leather subculture, which surfaced after the Second World War, from where it passed into heavy metal and glam rock and where, admittedly taken beyond parody, it has incorporated flashing lights or emitted sparks and even flames. But the codpiece may have made a return as much as two hundred years earlier than that – it depends on how you view the sporran . . .

Until the eighteenth century (that century again) the sporran, first cousin of the medieval belt pouch and, the Highland Scot's kilt being pocketless, an essential item of his ensemble, had been worn at the waist and to the side. Now, belatedly following the style of the Middle Ages, it was brought it to the front and dangled low. At this stage it retained its traditional characteristics: it was small, simple and made of leather; but it quickly followed the medieval pattern of increasing ostentation. By the next century it came in multifarious sizes, colours and materials, complete with swaying tassels and other decoration. Some were of animal hair – some even of whole small animals. Unlike the conventional codpiece, of course, the sporran remained uncoupled from the garment behind it: a semi-detached codpiece or, as Desmond Morris prefers, 'a surrogate pubic area'.

Whether regarded as the one or the other, the genital significance of the sporran was given further impetus when the kilt was adopted as Scottish army uniform – and it became obligatory to wear nothing underneath. In 1815, after the victory of Waterloo in the Napoleonic Wars and the subsequent occupation of Paris, the Emperor of Russia requested that men from each of the Highland regiments should parade before him at the Élysée Palace. As they stood to attention, the emperor fingered their kilts and sporrans as he passed them, but grew more inquisitive when he reached the gigantic figure of Cameron Sergeant Thomas Campbell, 'and had the curiosity to lift my kilt to my navel,' Campbell later wrote, 'so that he might not be deceived'. So equated in the Scottish military mind with manliness was 'going regimental' (as it was then termed; 'commando' in modern parlance) that even on the Western Front during the First World War, when officers or senior NCOs inspected their men, they did so with a mirror tied to the end of a golf club or walking stick to check orders were being obeyed. As late as the 1960s barrack room mirror-checks were common. Today, kilt wearers in the civilian population, just like those in the military, regard only those who hang free as true Scotsmen.

The polymathic novelist Anthony Burgess rather agreed – though actually, when he was teaching in Malaya, he wore a sarong:

> I carry a penis and a pair of testicles. These are not particularly handsome, unless stylised into the Holy Trinity or a Hindu lingam. They are inconvenient, and men's clothing is not well designed to accommodate them. I promise myself to declare Scottish ancestry and wear a kilt.
> (*The Real Life of Anthony Burgess*, Andrew Biswell)

The artist/sculptor Eric Gill was more vehemently anti-trousers. In a little book (*Trousers and the Most Precious Ornament*) published just before the Second World War he derided the garment, in which the penis was 'trapped all sideways, dishonoured, neglected, ridiculed and ridiculous – no longer the virile member'. Mind, Gill wasn't a codpiece man: he wanted a return to the medieval smock, which he wore himself with the precious ornament unencumbered by underwear. But Eldridge Cleaver, who was as one with Gill on the subject of trousers, which 'castrated' men ('the penis', he said with confused poetic charm, 'is withering on the vine'), was a codpiece man, and he was responsible for one of the two more-or-less serious attempts in the last decades to return the item to the mainstream.

A black American, Cleaver was variously jailbird, drug addict, Black Panther civil rights leader, revolutionary Marxist, author, born-again Christian, presidential candidate, radio talk-show host and environmentalist. And in the 1970s, passionate advocate of the codpiece. He produced a prototype, built into a pair of trousers ('Cleavers'), which, he said, would 'put sex back where it should be'. What he'd devised went far beyond the recognisable codpiece cup – indeed Cleaver's cup ranneth over, being, in effect, an external set of genitalia ('anatomically correct', he superfluously emphasised) that might have put the wearer in danger of arrest for indecency.

Cleaver's campaign withered on the vine. So did Jennifer Strait's twenty years later – and she not only had academic clout but also the Internet on which to rally penis-possessors to the flag.

Professor of apparel, merchandising and textiles at Washington State University, Strait launched her attempt to rehabilitate the codpiece – the conventional codpiece – because, in agreement with Cleaver in this respect, she thought men's clothing reflected 'absolute sexlessness'. And she sought sexual equality: 'If women exaggerate their breasts with bras, why shouldn't men enhance their body form?' Her campaign began cautiously, with a T-shirt ('Bring Back the Codpiece'), her intention being to introduce codpiece-sloganed boxer shorts, aprons and bumper stickers and only move on to codpiece production once a head of steam had been generated. Her expectations were deflated. The men's fashion magazines ran a mile from her approaches and T-shirt orders only trickled in. Penis-possessors weren't in the market for retro enhancement: they saw ridicule looming. After two years Strait gave up, leaving her website (last modified in 1995) a forlorn testament to a lost cause.

Both advocates of the codpiece had tried to catch the coattails of the increasing sexual frankness that characterised the decades leading up to the millennium. But the zeitgeist was beyond the peek-a-boo respect that the codpiece had once accorded the penis – in fact, the age has outed the penis. Penises of male strippers jiggle at hen parties in pubs, clubs and village halls to shrieks of 'Off, off, off!' On stage and screen (big, small and finally computer) real-life, flesh-and-blood penises leap like salmon. It seems quaint that forty years ago the hippie counter-culture musical *Hair*, the first theatrical production to feature full frontal male nudity, created a firestorm of protest – for a revival in 2008 theatre managements were more concerned to caution patrons that they might be exposed to herbal cigarette smoke from the stage than warning them about the sight of dangling bits and pieces. So many penises have been among the dramatis personae of so many plays that managements no longer bother to draw attention to the fact, though they do warn when strobe lighting is used. When Nicole Kidman appeared on the London stage in *The Blue Room* and was hailed as 'pure theatrical Viagra', no one seemed interested that her co-star was just as naked and turning cartwheels into the bargain – surely a theatrical first. So commonplace are penises on the boards – simulating masturbation, fellation, penetration or, as with a celebrity cast as himself, doing little more than coming on to milk the applause – that only a *coup de théâtre* will get a review; the penis that came to boiling point with a whistling kettle in *Buff* made inches of column space – if fewer than *Puppetry of the Penis*, which invites laughter with a display of genital origami. Long gone are the days when male nakedness on television was restricted to buttocks; in mainstream cinema even the erect penis and not necessarily only simulating can be sighted. Since Richard Gere revealed all in *American Gigolo* in the 1980s, it is almost de rigueur for screen actors to showcase their genitals (Kevin Costner was so put out when the studio cut a few frames from a movie in which he emerged from the shower, he threatened to sue). On the Internet's porn sites, tens of thousands of penises preen and prod at work and play –

possibly an oxymoron.

Penises have also become intrinsic to the age's iconography, in the tool kit of pop art and advertising. The Chapman brothers substitute penises for noses; a female artist shows her 'Wall of Wangers' in a London gallery, a display of eighty-eight penis-casts of eighty-eight different erections, making it seem that the Plastercasters pop groupies in the 1960s weren't really trying; in the ultimate in penis art, a portrait of former Australian prime minister John Howard is painted by an artist using his penis as the brush (one up on Renoir who said 'I paint with my prick', though he meant that figuratively and viscerally). Advertising turns to erections sparingly (always humorously, with covert cushion or towel or sometimes the product being promoted) but is addicted to the penile double entendre: 'one-and-a-half inches in the lunchbox' (a bigger packet of biscuits); '15½ inches and no wrinkles' (a shirt); 'Can you keep it up for a week?' (a national vegetarian campaign). Size does (or doesn't) matter matters almost as much to advertising as it does to penis-possessors.

The average penis-possessor is not likely to theorise whether the penis's status is diminished by frank flauntings or constant allusions. He does not feel a personal diminution. Indeed, in the years of his pomp he is subconsciously sure that his virile member is not only not dishonoured or ridiculous within his trousers but sallies forth with him, on display in some virtual way that cannot be defined by mere words. The gait of some penis-possessors leaves no doubt of this cocksureness, involving their thighs in an apparently necessary circumnavigation of what lies between them – something, it seems, of an order preventing simple forward locomotion: a codpiece worn interiorly, perhaps, and possibly a poulaine. Such penis-possessors tend to sit on public transport with legs flung wide, as if bringing their knees closer together simply can't be done. So prevalent is the habit among the Japanese – an irony here, given the Japanese's low standing in the international measurement tables – that on the Tokyo underground the carriages have 'do not . . . ' stickers on the windows. In her contribution to the book asking women how they would behave if they had a *Dick for a Day*, Maryanne Denver wrote: 'I would stick it on my forehead and parade around the way regular owners do.' She was expressing herself metaphorically; but where some men are concerned, only just.

1. During the First World War the Italian Prime Minister Vittorio Emanuele Orlando wore a fascinum on a bracelet to ensure victory for the Allies – a residual belief in penis power, perhaps, or just a case of covering all the bases.

   Today in certain cultures men have similar phallic amulets. In Thailand, one or more are worn hanging near the penis from an intricately woven cord around the waist under the clothes, to absorb any negative energies directed by others to their genitals, and to increase their sexual attractiveness (and maybe bring gambling luck too). The Thai name is *palad khik* – 'honourable surrogate penis'.

2. A male baboon says hello to another by pulling on his penis, the courtesy being reciprocated; proto-human man almost certainly did likewise. The Walbiri of central Australia today hold the penis of a visitor as the equivalent of shaking hands.

   A hangover from the practice of genital oath-taking still exists in rural areas of Mediterranean countries, men clutching or touching themselves when emphasising the veracity of what they are saying, or to avert bad luck.

3. The primary biological function of the penis is to deliver semen to the vagina, to accomplish fertilisation – a function, however, that occurs in negligible ratio to its use purely for pleasure and even less in expelling urine 'from the bladder to the porcelain of the outside world', as John Gordon phrases it in *The Alarming History of Sex*. That semen travels the same penile piping (though not, of course, at the same time) Gordon regarded as an example of 'nature's recurrent economy'; an obverse view might be that when it comes to genitalia, God is not a sanitary engineer.

   Women are exasperated by men's cavalier attitude to urination (as a character in Laurie Graham's novel *The Ten O'Clock Horses* says, 'a man having a jimmy couldn't aim it down the Mersey Tunnel'), but the German psychoanalyst Karen Horney over fifty years ago believed it indicated 'fantasies of omnipotence, especially those of a sadistic character'.

   Modern Swedish women would seem to agree. In 2000 they demanded that men use the toilet sitting down, partly for hygiene reasons but more crucially because standing up was deemed to be 'triumphing in [their] masculinity and therefore degrading women'. Feminists at Stockholm University campaigned to scrap the campus urinals, and primary schools began to get rid of the wall-fixed porcelain to acclimatise young males to the new order.

The campaign has spread to Germany where, a survey suggests, 40 per cent of men now sit, the same percentage as in Japan.

4. A variation in the seventeenth century was that conception was the result of magnetic energy produced by the friction of intercourse, the female reproductive tract being 'magnetised' by the male 'spark'. In the middle of the following century the word 'spunk', the principal meaning of which is courage, became a colloquialism for the safety match – which requires friction to spark, leading, two stages removed, to its slang use for semen. 'Mettle', also slang for semen, derived from it having the same proper meaning as spunk.

5. Yin–yang is a dominant concept in Chinese philosophy, representing the two primal cosmic forces in the universe. Yin (moon) is the receptive, passive, cold, female force, yang (sun) the active, hot, masculine one. Summer and winter, night and day, health and disease, woman and man – all things follow the principles of yin–yang and are in some way related to each other.

6. Japan, an overtly phallic culture into the twentieth century, still celebrates fertility festivals that give a good idea of what went on in medieval Europe. Men carry a giant wooden (or, increasingly, pink plastic) penis in procession to a local shrine, with many wearing papier mâché penises on the front of their costumes, and woman cradling wooden penises; phallic ice creams, lollipops and other snacks are on sale to the crowds. At some festivals a giant phallus of straw is set fire and plunged into a straw vulva, as milk-white sake is spattered about. Many festivals have disappeared in recent times – again, Westernisation.

7. When Rasputin was murdered in 1916 by a group of nobles fearful of his influence over Tsarina Alexandra, they hacked off his penis. What happened to it during the next half century is unaccounted for but, according to Patte Barham who helped Rasputin's daughter write her biography, it was kept in a velvet box from which in 1968 her co-author's Parisian maid produced it, looking 'like a blackened,

over-ripe banana, about a foot long'. Nothing more was heard of it until the Russian museum of erotica opened in St Petersburg in 2004 with Rasputin's alleged organ as its prime exhibit.

8. Unlike many big-penised men, the affable comedian/actor Milton Berle was modest about his appendage, though Sammy Davis Junior once confirmed 'even down it's world class'. Asked by Davis how big his penis got erect, Berle replied, 'I don't know, I always black out first' – bringing to mind the observation of another actor/comedian, Robin Williams, that 'God gave every man a brain and a penis but only enough blood to make one work at a time.'

9. For centuries men have enhanced the appearance of their penis with inserts, but the practice has limited appeal in the West. It has become more popular since the 1970s, mostly in the gay community.

    Some twenty cultures of South East Asia traditionally accessorise their manhood. In India and Burma, many men sew little bells, some the size of a small chicken egg, under the skin of the shaft. The Malays, Koreans and Filipinos favour metal balls about the size of a hazelnut, the Sumatrans small stones, and the Japanese pearls. Three to a dozen inserts are common in all cultures – the Japanese *yakuza* (mafia) can have many more, sewing in a pearl for each year spent in prison.

    In the past, Asian royalty would remove one such accessory to bestow on a person deserving great honour.

    On the Indian subcontinent and in South East Asia generally, men have always attached jewellery through the penis itself, to enhance lovemaking: in India, a 'barbell', the *apadravya*, is inserted vertically through the glans; in Borneo and Sarawak, the similar *ampallang*, but this is fixed horizontally and sometimes intercepts the urethra. A combination of *apadravya* and *ampallang* is referred to as 'the magic cross'.

    Traditionally, the taste in North Africa and the Middle East has not been for penile but for scrotal adornment: the *hafada* ring is fastened anywhere through loose skin; multiple piercings are not uncommon, particularly the 'frenulum ladder'.

    Apadravya, *ampallang* and *hafada* augmentation is found in the West, but the most popular is the so-called 'Prince Albert' ring, which pierces the underneath of the penis behind the glans, passes up through the urethra, and exits at the tip. That Queen Victoria's consort wore a 'dressing' ring like this is widely believed but is an urban myth emanating from the 1970s.

PART THREE

**UNEASY LIES
THE HEAD THAT
WEARS THE
CORONA**

Mankind is ruled by the Fates, they even govern those private
parts that our clothes conceal. If your stars go against you
the fantastic size of your cock will get you precisely nowhere . . .

Juvenal

HAZARDS OF OWNERSHIP

Pluses and minuses

One night in the summer of 415 BC, just before the Athenian army set sail to wage an unpopular war against Sicily, someone knocked the penises off the herms across the city – hundreds of penises, in public places, in courtyards, in the doorways of private houses. According to the nineteenth-century English classical historian George Grote, men waking up to find their phallic guardians castrated felt as if Athens 'had become as it were godless'.

It was never discovered who carried out the deed, or why. In the 1990s the feminist Eva Keuls, professor of classics at Minnesota University, asserted in *The Reign of the Phallus* that the perpetrators were a group of women making an antiwar protest and a protest against their phallocentric world into the bargain. Other historians have dismissed her claims as codswallop (etymology uncertain but in the context of this book identifiably synonymic). The story, however, emphasises the alliterative observation made by Gay Talese (*Thy Neighbor's*

Wife) that penises are 'very vulnerable even when made of stone, and the museums of the world are filled with Herculean figures brandishing penises that are chipped, clipped or completely chopped off'. The British Museum has a cabinet entirely filled with such dispossessions, hacked away in the name of religion by early Christians and divers fundamentalists in later periods such as the English Reformation and the French Revolution, though Victorian prudes made some contribution, taking hammers to the private parts of public statuary as an outrage against decency.[*]

The vulnerability of stone penises hardly compares to that of flesh and blood penises, and is as nothing compared to flesh and blood testicles. The penises of most mammals are protected within a sheath, from which they emerge only when engorged. The penises of human males, and their fellow primates, have no such sheath because, instead of being attached to the abdomen along their length, they are pendulant: they hang free. As for testicles, those of all mammals originate inside the body and in many species stay there; in some, they emerge only during the breeding season and then go back inside for safe keeping. But in others, including man, testicles descend into the scrotal sac before birth and here they remain through life. Given that the internal arrangement self-evidently offers maximum protection, it seems counterintuitive that the external arrangement exists, particularly, you might think, among the advanced higher primates. The reason that it does, a report in the *Journal of Zoology* suggested, is due to evolutionary locomotion. Species with a generally gentle way of getting about (elephants to moles) keep their testicles inside the skeletal structure. Species that run and jump (deer, kangaroos, horses, primates) have theirs outside – a development that occurred because their

---

'concussive' types of movement squeezed their testicles when they were internal and, since the reproductive tract has no sphincter, involuntarily expelled sperm, thus wasting it.

There are advantages to having external genitalia. Penises not in a rigidly fixed position have copulatory (and urinary) flexibility; external testicles house sperm in mildly refrigerated conditions, which keeps them lively and eager for the off (see Part 4, 'The "Precious Substance" Revisited'). But there are disadvantages: the increased danger of accident or assault, compounded in man by his upright gait, which points his delicate extremities to the front. His testicles are particularly vulnerable to injury and, being much more delicate than his penis – they are in fact part of his internal viscera – a blow or pressure to them can cause nausea or even loss of consciousness.

It's estimated that one adolescent boy in ten has been kicked in the testicles, with varying degrees of distress and consequence. Penises also suffer damage; small boys regularly turn up in A&E after the toilet seat has fallen on theirs, if not a sash window as happened to the infant Tristram in *Tristram Shandy*, giving a servant the opportunity to quip: 'Well, nothing in the Shandy household is well-hung.' Males of all ages regularly appear in A&E too, having carelessly trapped themselves in their zipper. Every year four million men in the UK and nineteen million men in the US sustain genital injuries taking part in sport and exercise. Some other kinds of accidents seem hilarious to anyone but the sufferer. A British holidaymaker in Fiji in 2005 dozed by a rock pool and awoke to find a barnacle clamped to his penis and had to be rushed to hospital. In 2006 a Croatian sat in a deckchair on his local beach but when he tried to get up found that his testicles were stuck between the slats – they'd shrunk while he skinny-dipped in the sea but had expanded to normal in the sunshine; an attendant had to cut the chair into pieces to free him.

Even having sex can cause a man physical injury. Perhaps a majority of men don't know they have a frenulum (if they're circumcised, they haven't), the thin elastic strip of skin that anchors the tip of the penis to the shaft on the underside. Frenulum means 'little bowstring' in Latin, and like a bowstring it can snap. Too vigorous sex is to blame and the injury is usually sustained by very young men or by men of any age during one-night stands, when the sex is more than usually vigorous. And there can be a lot of blood (Suzi Godson, *The Times* sex expert, wrote of an encounter in a hotel in which a man's frenulum ripped 'and when they left the room it looked as if someone had been murdered on the bed'). Besides being excruciatingly painful, the injury almost certainly involves an acutely embarrassing visit to A&E for some needlework, but it mends quickly. A broken penis is a different matter.

Every year at least two hundred Americans, and thirty to forty Britons break their erect penis. A few, bizarrely, do so by tucking themselves into their underwear while still tumescent, but almost all do it during violent intercourse. A few, equally bizarrely, do it by falling out of bed during this and buckling their erection on the floor; a few by being too physical and crumpling against their partner's pubic bone or perineum; but most do it when their partner is on top in the riding position and rotating her hips.

The lining of the penis, the tunica albuginea, is about the thickness of thin cardboard and protects the spongy chambers that fill with blood during erection, and it has a safety factor of ten times the normal erect-state pressure. If that is exceeded, however, the tunica breaks – and breaks with an audible crack.

Again the pain is excruciating, the penis swells and turns the colour of a ripe Victoria plum, and surgical repair is needed, as well as six weeks' bed rest with the penis in splints. The break will heal, though the likelihood is a penis that kinks on subsequent elevation. If the break is extensive, the unfortunate penis-possessor can develop Peyronie's disease in which fibrous plaque accumulates at the site, erection becomes painful and rigidity diminishes – and the penis can take a sharp left or right turn and be grossly misshapen, bulging at top and bottom but seeming to have an invisible clamp around the middle. Four in ten men with Peyronie's have a permanent degree of erectile dysfunction.

Unlike a torn frenulum, Peyronie's (which can also have medical causes) is suffered mostly by middle-aged men, though the prevalence and incidence are hard to establish; many don't seek help because of their embarrassment. Current literature suggests 3–9 per cent of men are victims.

The dispossessed

According to Rabelais in *Gargantua and Pantagruel*, the first part of his body that early man protected (with tough fig leaves) was 'the staff of love and packet of marriage', words spoken by Lady Humphrey de Merville in exhorting her husband, on his way to war, to attach a jousting helmet to his groin. It was in Rabelais' century, the sixteenth, that new technology allowed the 'genitall shield' to be incorporated in suits of armour: no less boastful in size than the codpiece but serving a more critical function. When purpose-made protection is lacking in dangerous circumstances, men take emergency measures. Troops being carried in aircraft that came under fire from the ground during the Second World War often chose to sit on their helmets, protecting their lower head rather than their upper.

In the aftermath of wars, penises can be even more vulnerable than during them: the victorious have a habit of emasculating the vanquished. Once, it was universally believed that making enemies incomplete in their parts would prevent them from getting to the other world, from where they might wreak vengeance; the belief still seems to exist in parts of the Mediterranean and Middle East. Mostly, of course, penis-possessors at war have castrated other penis-possessors, to take from them the very thing that makes them men, their manhood, and in so doing, 'feminise' them.

The Egyptians, Babylonians, Hebrews and Ethiopians among other nations regarded penises as trophies of war, with bookkeeping scrupulousness: after invading Libya in 1200 BC, the Egyptians went home with a haul of 13,240 (six Libyan generals, 6,359 Libyans, 6,111 Greeks, 542 Etruscans, 222 Sicilians). The Aztecs weren't trophy collectors, preferring to string taken penises along the road to shame their foes – as the Spanish found out when they invaded central Mexico in the sixteenth century. Whether taken as trophies or not, the cutting off of enemies' genitalia has occurred in every kind of conflict: the Normans, for example, castrated Englishmen in the aftermath of the Battle of Hastings, including the already dead King Harold who, William of Poitiers relates, had his 'leg' cut off – leg being a Norman euphemism; the English and French castrated each other on the famous battlefields of the Hundred Years War; the remnants of Napoleon's army in the retreat from Moscow, starving in the snowy countryside, were hunted down by the Cossacks and castrated in their hundreds. The conflicts of the twentieth and twenty-first centuries, big and small, have not been exempt.

Penis-possessors haven't always needed the excuse of war to deprive other penis-possessors of their manhood – in the years of the codpiece the Turks waylaid Western travellers to see if the contents lived up to the packaging and removed what they found when it did not, out of indignation and perhaps pent-up relief that it did not. Fear of the potency or the size of the penises of other races has historically led to instances of castration. The soldiers of the Roman emperor Hadrian cut off the organs of living Israelites – because circumcision, a religious rite of Judaism, permanently exposed the glans, as occurs in the uncircumcised when the penis is erect, thereby giving them the reputation of being pathologically lustful – flinging them heavenwards and taunting God: 'Is this what you have chosen?'[2] Bible Belt America, which believed that the black man was the descendant of Noah's accursed son Ham and had 'the flesh of asses', during the second half of the nineteenth century and the first half of the twentieth lynched over four thousand for the alleged rape of white women, almost all of the men first having their genitals cut off.

Castration has always been in the repertoire of executioners and torturers. During the Crusades (when Christians and Muslims emasculated each other with equal enthusiasm) the French knight Thomas de Coucy hung up his captives by their genitals until they ripped from their bodies; in his autobiographical memoirs (De vita sua) Guibert de Nogent gives an account of this, and the nightmares it gave him. Centuries later the Inquisition similarly suspended some of those unlucky enough to come to its attention – and targeted the penises of others with the 'crocodile shears', a metal contraption of hemicylinders with internal teeth, which was heated red hot before being clamped in place. Castration was also an aspect of hanging, drawing and quartering, which for five hundred years – until as late as the eighteenth century – the English meted out for high treason (often liberally interpreted: Henry VIII executed a few obstinate Catholic clergymen to bring others into line over his reform of the Church; he dispensed with an alleged lover of his fifth wife in this way too). Victims were half hanged, their 'privy members' then cut off and held before their face while they could feel the agony and humiliation, before these were thrown into the fire, followed by the bowels from their sliced-open body – a fate suffered by Charles I.

Castration of the testicles was commonly one of the prices for a variety of crimes in the whole of medieval Europe, including for counterfeiting coins and for poaching royal deer; in France during the Protestant Reformation it was the penalty for homosexuality, which the convicted on balance must have thought preferable to death, commonly the penalty in different cultures and periods. But a second offence meant the loss of the penis and a third burning at the stake. Across Europe, for rape, for the taking of the virginity of a daughter of a peer, and, in some parts, for adultery with another man's wife, the penis and the testicles were forfeit – as in China, Japan and India (where a man who seduced the wife of his guru was made to sit on a hot plate and then chop off his penis himself). European revenge narratives cite many instances, the most famous that of the French philosopher Pierre Abelard who fell in love with his pupil Heloise. After she became pregnant, her uncle and a gang of kinsmen tracked him down and, he later wrote, hacked off 'those parts of my body by which I committed what they complained about'.[3] Clerics and monks guilty of sexual transgressions were often separated from those selfsame parts. A lay brother who impregnated a teenage nun of the Gilbertine order at Watton in Yorkshire was lured by her fellow sisters to their monastery, where she was forced to castrate him before returning to her cell.

Abuse and betrayal have undeniably always driven some women to castrate men, needing no coercion to wield the knife. But such handiwork became a worldwide phenomenon from the early 1990s after John Wayne Bobbitt, a small-town former US Marine, had his penis cut off by his wife Lorena. Across America, and from China to Peru, copycat cases began to occur, with Thailand becoming the epicentre: by the end of the millennium, over a hundred cases had been reported to Thai police, who admitted there were probably many more but the victims preferred to keep their loss to themselves. Penises, and testicles, can of course be reattached and even returned to normal functioning – if, that is, they can be found. Bobbitt was lucky: his wife had thrown his penis over a hedge and it was recovered. A man in Alaska was equally lucky: his partner had flushed his down the toilet but it turned up at the local waterworks. In thirty-one of the above Thai cases Bangkok Hospital was able to give another meaning to 'friends reunited'. Other severed penises, however, had gone for ever – women had fed them to their ducks or chickens or put them in a blender or down the waste disposal. One man in India had to wave goodbye to his penis after his wife attached it to a helium balloon.

Female emasculation of men is generally a singular activity but a collective one occurs in Emile Zola's *Germinal*, though the shopkeeper Maigrat, guilty of sexually harassing or raping many of his female creditors, has already expired in a fall from a roof during a miners' strike:

> ... the women had other scores to settle. They sniffed around him like she-wolves, trying to think of some outrage, some obscenity to relieve their feelings.
>
> The shrill voice of Ma Brûlé was heard:
>
> 'Doctor him like a tomcat!'
>
> 'Yes, yes, like a cat! The dirty old sod has done it once too often!'
>
> Mouquette was already undoing his trousers and pulling them down, helped by la Levaque who lifted the legs. And Ma Brûlé, with her withered old hands, parted his naked thighs and grasped his dead virility . . . and pulled so hard that she strained her skinny back . . . The soft skin resisted and she had to try again, but she managed in the end to pull away the lump of hairy, bleeding flesh...

But such a collective act is not confined to a novelist's imagination, or to the deceased. In Cambodia, women dragged a man arrested for a series of rapes out of the police station, cut off his penis, put it through a meat mincer and then made him eat it.

Castration is a messy business, however done; few women resort to it. Rather more women who don't, but who nonetheless think that only genital retribution will satisfy their grievance, adopt a hands-off approach – they administer scalding water or hot fat. Yet removing or maiming a man's penis may not always be a vengeful act. A Beijing housewife had nothing but love in her heart when in 1993 she de-penised her spouse with a pair of scissors. A fortune teller had told her that his inadequate organ was the problem in their relationship. She snipped it off in the hope of making it grow back bigger and better . . .

Other dispossessions

Men have castrated other men for more reasons[4] than bloodlust – principally to provide servants, guards, administrators and priests. The Carib Indians (who gave their name to the Caribbean) castrated boys captured from their enemies for culinary purposes. Removal of a male's testicles before puberty prevents the hormonal rush into adulthood. A cannibal people, the Carib appreciated that castrates' flesh remained unmuscular and therefore tender until such times as they went into the pot.

East and West, castrated prisoners and criminals were the first servants, but demand for them exceeded supply and, as boys castrated before puberty proved more docile and trustworthy, slave traders saw the business opportunity and shipped in boys from other countries – the prettiest, it has to be said, fetching up in male brothels. Most of the eunuch class that came out of Africa, through Egypt and the Sudan, were 'fully shaved' – they were deprived of penis and testicles. Only such males were allowed in the harems of the Turkish Ottoman sultans. Elsewhere in the palaces, eunuchs usually shorn only of their testicles were employed, and they were white, not black: the Ottoman Empire at its height spread out of Asia into parts of eastern Europe and considerable numbers of eunuchs came from Hungary, the Slav lands, Germany, Armenia, Georgia and the northern Caucasus. All eunuchs in imperial China were fully shaved – emperors were ever fearful that a rival dynasty might be founded and took no chances with an enemy within. More secure in their power, the Moguls in India allowed all their eunuchs to retain their penis.

Castrating young boys was simple: pressure on the carotid artery rendered them unconscious, after which their testicles were crushed, often by hand, permanently damaging the seminal glands; alternatively their testicles were strangulated with cord so that they became necrotic and fell off – farmers castrate lambs in much the same way with elastic bands.

The adult scrotum and testes were forfeit to the knife.

Cutting off the testicles alone didn't usually endanger life. But cutting off the penis did and fewer than one man in five survived African castration; even in in the Sudanese town of Tewasheh, once one of the world's largest suppliers of eunuchs, only three thousand of the thirty thousand castrated annually didn't die. The procedure was crude. The man was strapped down, his genitals bound with yarn to stop the circulation and then sliced away, the wound cauterised with a hot iron or tar and a bamboo rod inserted to keep the urethra open. Next the castrate was buried up to his navel in sand or mud and given nothing to drink for five or six days. If then his urine flowed, he stood a chance of living; if not, he'd die an agonising death, unable to empty his bladder – if, that is, he didn't die from loss of blood or septicaemia. By contrast, the Chinese washed a man's genitals in hot pepper-water to desensitise them, then removed them with a curved blade dipped in antiseptic lime juice. The wound was sealed with a silver plug and he was walked around for several hours before being allowed to rest. Three days later, the plug was removed. The castration procedure was so radical that if the castrate lived (perhaps half did) he would never be able to urinate standing up other than through a quill.

Penis-possessors may be disbelieving, but there was no shortage of willing candidates to become eunuchs in China, India and Byzantium, the breakaway eastern half of the Roman Empire that fell to the Ottoman Turks in the fifteenth century: preferable to be inside the palace walls without some or all of your compendium than outside with it intact, but in abject poverty. Wealth and opportunity beckoned within and parents with several sons would often have one or two castrated in the hope of getting them into service; in 1644 there were twenty thousand applications for three thousand vacancies in China's Forbidden City, which at the time employed seventy thousand eunuchs. It was possible for a eunuch to rise to administrator, military commander or even to confidential adviser, and in Byzantium eunuchs were so well regarded for their supposed incorruptibility that eight of the chief posts in the empire were specifically reserved for them.

The Ottoman and Chinese dominions collapsed in the early twentieth century and with them the era of the eunuch. The last in the Forbidden City streamed out in 1912, each carrying a pottery jar containing his severed organs (known in euphemistic Mandarin as 'the precious treasure') preserved in alcohol, to be interred with him when he died so that when reborn he would be whole again.

From at least the ninth century AD eunuchs sang in Christian Byzantine choirs. The larynx of men deprived of their testicles, like the rest of their body, did not grow normally and their voice retained a boy's vocal range while at the same time developing extraordinary power. When Italians began to experiment with complex polyphonic choir scores in the sixteenth century, choirmasters, forbidden by papal degree to recruit women, stealthily introduced castrati. Castration was illegal (but not unknown: poor Italian mothers sometimes had a son castrated to sell to Turkish traders, who paid good prices). Hardly surprisingly all those who tried for choir places had not been to the local barber to be given opium, placed in a tub of very hot water and rendered almost unconscious and then have his parts sheared away; no, all of them had met with a tragic 'accident'.

There was an upside to prepubescent testicular removal. The castrati did not go bald and, according to modern statistical research, lived thirteen years longer than average; and the seventeenth century made some the stars of the Italian operatic stage ('Long live the knife!' shouted the adoring crowds), the greatest of them becoming seriously rich. Women threw themselves at the castrati and even if tales of their conquests are exaggerated, some did have a lot of sex – losing testicles does not mean losing the ability to get erections and even to ejaculate. What Juvenal wrote about Roman matrons and young girls was true of Italian ladies: 'They adore unmanly eunuchs – so smooth, so beardless to kiss, and no worry about abortions!'

But there was a downside, besides the inability to father offspring. Hormonal imbalances meant womanly breasts and hips, a weak bladder, poor eyesight, a lack of manly body hair and often an unusually small head; and for many a condition called macroskelia, which made the bones of the ribcage, legs and arms continue to grow – the arms of some castrati reached to their knees. Onstage the castrati were usually a head taller than anyone else, which was awkward for those who played women's roles. The era of the operatic castrati ended in the early nineteenth century with a change in musical tastes and the rise of the diva, by which time an estimated four to five thousand boys a year had experienced 'accidents'. But the last castrato did not leave the Sistine Chapel choir until 1913 – a succession of popes continued to turn a blind eye for the glory of God.

There have always been men prepared to be castrated for their religious beliefs. The priests of many civilisations were, including those of the Roman Cybele cult whose novitiates castrated themselves on the yearly public 'day of blood'. What is extraordinary is that as the flutes played and the drums beat, some spectators, presumably normal penis-possessors, got so carried away that they joined in. According to James Frazer (*The Golden Bough*) 'man after man, his veins throbbing with the music, his eyes fascinated by the sight of the streaming blood, flung his garments from him, leaped forth with a shout, and seizing one of the swords which stood ready for the purpose, castrated himself on the spot'.

Early Christians were obsessed with self-castration as the ultimate means to chastity ('and there be eunuchs who have made themselves eunuchs for the kingdom of heaven's sake' – Matthew 19:12), one being the third-century theologian Origen of Alexandria. Fifteen hundred years later the Russian Skoptzis, who broke away from the Russian Orthodox Church, followed the same biblical reading, the sect surviving into the twentieth century. The Karamojong of northern Uganda and the holy sadhus, still going strong in India and Nepal, cut nothing off but have a unique way of putting the penis out of commission; from an early age they hang heavy weights from it so that eventually it becomes two or three feet long and can be tied up in a knot – some sadhus carry theirs about in a cloth basket.

Extreme religiosity still accounts for a proportion of those in the West who take a knife to their genitals, in any combination of one testicle, two testicles, or penis, or *tout ensemble*. Drunks show up at hospital from time to time having removed a testicle or even two for no better reason than it seemed like a good idea at the time, frequently involving a bet. Those who fully castrate themselves are usually mentally disturbed or desperate transsexuals convinced they're in the wrong anatomical body.[5, 6]

The medical profession over centuries has been keen on removing a man's testicles as prevention or cure for many conditions. Castration was a regular treatment for hernias in the Middle Ages (as was piercing the testicles of those with bubonic plague). French physicians routinely castrated patients suffering from leprosy, rheumatism or gout. Doctors have been reluctant to relinquish the idea that the testicles play a part in unrelated bodily ills and abnormalities – as latterly as the early twentieth century in America epileptics, alcoholics, the insane, homosexuals and habitual masturbators – this last had been strongly advocated in Victorian England – were castrated.

Half a century ago it was especially unfortunate to have a micropenis; the most common way of dealing with the worst cases (almost no penile shaft, the glans virtually sitting on the pubic skin) was gender reassignment: a boy's testes and vestigial penis were taken away, an artificial vagina fashioned, and the sufferer was told he was now a girl and prescribed female hormones for the rest of 'her' life. Today, advanced surgery can increase the abnormally small organ to something like normal dimensions, using muscles from the forearm, without loss of erogenous sensation. Men thus transformed, as the *New Scientist* reported in 2004, are now able 'to enjoy a full sex life and urinate standing, some for the first time'.

THE NEUROTIC PENIS

All in the mind

Can a man's penis be stolen from him by sorcery? Once, such a fear seems to have been universal. It looms large in the folktales of preliterate societies. It appears in ancient Chinese medical texts. It was part of the European medieval mindscape, resulting in hundreds of witches being burnt at the stake for penis theft. In the *Malleus Maleficarum*, the fifteenth-century guidebook on witches and their ways, the German Dominican monk and witchfinder Jacob Sprenger asserted that witches

> collect male organs in great numbers, as many as 20 or 30 members together, and put them in a bird's nest, or shut them up in a box, where they move themselves like living members and eat corn and oats, as has been seen by many and is a matter of common report . . . a certain man tells that, when he had lost his member, he approached a known witch to ask her to restore it to him. She told the afflicted man to climb a certain tree, and that he might take which

he liked out of a nest in which there were several members.
And when he tried to take a big one, the witch said 'You
must not take that one, it belongs to a parish priest.'

Such was the pathological state of medieval thinking that
witches were believed to have intercourse with the devil,
whose penis was said by some to be at the rear and covered
in scales, while others declared that it was forked or, indeed,
that he had two, in either case for the purpose of simultaneous
front and rear penetration. The devil's ejaculate was reputed
as cold as ice and exceeded that of a thousand men. Freud
believed that the witch's broomstick was really a metaphor for
'the great Lord Penis'.

In the modern world, the delusional disorder of penis
theft is largely confined to the countries of West and Central
Africa. Those from Malaysia, Borneo and southern India and
China have a related but different anxiety – the belief that it's
possible for their penis to shrink into their abdomen, which
will cause them to die and become ghosts. Periodic epidemics of
hysteria occur, sweeping through towns or cities or even entire
countries. In a recent isolated instance in the Sudanese capital
Khartoum, a rumour warned men not to shake hands with a
'mysterious West African', whose handshake melted genitals.
Scores who thought themselves afflicted sought medical
treatment. Alleged penis thieves in Africa are routinely hanged
or set alight with petrol by angry mobs. But the mysterious
West African was never found, which isn't surprising: the
rumour was a hoax – spread by text-messaging.

Anthropologists call penis panic a culture-bound syndrome;
South East Asians call it *kora*: 'head of the turtle' in Malay. And
from time to time *kora* packs outpatient departments with
terrified men, as it did in Singapore during an unaccustomed
spell of very cold weather in 1967. Cold naturally tightens the

genitals. Hundreds of men were otherwise convinced about the cause of their shrunken parts and headed for the hospital, most with a hand in a trouser pocket or under a dhoti, others with a string disappearing under their clothes (some with a rock tied to the string), with a few unashamedly hanging on to their exposed member with both hands, those taking no chances having a safety pin thrust through their glans. During possibly the largest outbreak of *kora* paranoia in modern times, which occurred in 1984–5, five thousand men in the Chinese Guangdong province used similar preventative measures, though, with cultural nicety, some chose to prevent fatal retraction by holding their penises with chopsticks.

Venturing into the unknown

Men want sex but from the beginning of time have been in trepidation of its source. Long before Freud and psychoanalytical theory, primitive man thought that a woman's external sexual organs looked like the site of male castration – and that such a fate might befall him if he ventured in.

The vagina was a place of procreative darkness, a sinister place from which blood periodically seeped as if from a wound. During menstruation, counselled the medieval *De secretis muilierum* (On Women's Secrets), a woman was so full of poison that a man who had intercourse with her could be made leprous or cancerous. The vagina was so evil, according to Muslim thought, that a man who looked into its entrance could be turned blind. First crossing its threshold was dangerous, men everywhere believed. In many countries and still in parts of Africa and India, the risk of deflowering a bride came to be taken for the bridegroom by an older man, village headman, or feudal lord or cleric, whose powerful status supposedly could overcome the malevolence within – a belief undoubtedly encouraged by older men, village headsmen, feudal lords and clerics. In the East, high-caste males sometimes commanded a slave – who was expendable if the task proved injurious – to see to the matter.

In some countries where this defensive strategy was common, a parallel practice was for alpha males to take 'the first fruits of the bride' on her wedding night. Over four thousand years ago, according to the Epic of Gilgamesh, the people of Uruk (modern-day Iraq) were less than happy about their king's insistence on 'the king to be first and the husband to follow'.

The Greek historian Herodotus believed the custom was unique among an obscure tribe in Libya, but it was widespread in the ancient world – and instances of it occurred here and there almost to the present day, notably in the Ottoman Empire, the centre of East–West interactions for six centuries up to the early twentieth century, and in western Armenia, where Kurdish chieftains reserved the right to bed a bride on her wedding night.

A considerable body of writing maintains that the right of the first night (*jus primae noctis*) was practised in medieval Europe, but there is no incontrovertible evidence (blame Voltaire for the widespread currency of the claim; in the eighteenth century he took it to be historically authentic). There is, however, evidence that a liege lord had the right to lie on a bride's bed and pass his naked leg over her (*droit de jambage*); in some places the bride was obliged to make a payment to him ('legewite' in Anglo-Saxon law, a 'tax for lying down') in compensation for the loss of her (assumed) virginity to another. The custom was symbolic, less to do with sex than a display of lordly power over the peasantry – and an interesting take, perhaps, on the idiomatic phrase 'leg over'.

Even when made safe, men feared that the vagina, already attributed mysterious sexual power (did it not conjure up a man's organ, absorb it, milk it, spit it out limp?), might be insatiable. 'Three things are insatiable,' runs a Muslim aphorism, 'the desert, the grave, and a woman's vulva.' The Arab world was victim to the most rabid fear of the voracious vagina. 'I saw her vulva!' laments a lover in the fifteenth-century masterwork *The Perfumed Garden*. 'It opened like that of a mare at the approach of a stallion.' The author warned readers: 'Certain vulvas, wild with desire and lust, throw themselves upon the approaching member.' Europe had much the same anxiety about insatiability, which in some women, it was said, caused the clitoris to rear up to the length of the male member. 'Though they be weaker vessels,' wrote the Elizabethan musician Thomas Whythorne, 'yet they will cover two, three or four men in the satisfying of their carnal appetites.'

Prior to intercourse, Thomas Bartholin declared in a popular seventeenth-century medical treatise, the vagina 'gapes to receive the Yard, as a Beast gapes for its Food'. Beast – or, four hundred years later, as a ravening bird in Lawrence's *Lady Chatterley*, Mellors telling Connie about intercourse with his wife: ' . . . and she'd sort of tear at me down there as if it was a beak tearing at me. By God, you think a woman's soft down there, like a fig. But I tell you the old rampers have beaks between their legs'. Or worse: teeth. The Middle Ages believed that some witches, with the help of the moon and magic spells, could grow vaginal teeth with which to rend men; in the myths and legends of many nations from China to North America but especially in South East Asia, vaginas with teeth, or even weapons, castrate or kill their sexual partners.[7]

A lesser but no less real anxiety for some men was and remains that the vagina will not let them go. In *The Second Sex*, Simone de Beauvoir wrote understandingly of young men's nervousness in venturing 'into the secret dark of women, once more feeling childhood terror at the threshold of a cave or tomb', and their fear that 'the swollen penis might be caught in its mucous sheath'. While this is generally only a fear born of inexperience, entrapment of the penis – penis captivus – more often than not dismissed as an urban myth, can occur. In rare cases the twin levatores ani muscles on either side of the vagina can go into spasm so severe that it is impossible for a man to withdraw or for his penis to detumesce. In 1980 the *British Medical Journal* carried considerable correspondence testifying to experience of treating penis captivus, including a letter from a doctor who remembered as a houseman at the Royal Isle of Wight County Hospital seeing a young honeymoon couple being carried in on a single stretcher. Such an embarrassing situation is believed to have involved Lady Edwina Mountbatten and her black lover, the cabaret star Leslie ('Hutch') Hutchinson (who, a maudlin Lord Louis Mountbatten told the bandleader in the nightclub in which he was getting drunk, 'has a prick like a tree trunk and is fucking my wife'), in a London hotel in the 1930s; they were discreetly whisked away for medical disengagement.

Attracting the Venom

Men's anxieties about their sexual encounters have never ended with withdrawal. The spectre of venereal disease (the Ancient Egyptians termed it 'copulatory sickness') has always peered over their shoulder. Gonorrhoea was common in Europe from before the Middle Ages. But things worsened ominously in the last decade of the fifteenth century: a virulent strain of venereal disease, either brought back by Columbus's sailors from the New World or a mutated form of something that already existed (it is disputed), broke out like a plague.

Within days of infection, caused we now know by the spirochete bacterium, a small circular ulcer called a chancre appeared on the penis. Within weeks, bags of pus erupted all over the body from head to knees; the flesh fell from faces. The great pox (to distinguish it from smallpox, with which it shared some characteristics) quickly killed thousands – men and women both, though men blamed women, who, they thought, not only infected them but infected themselves. It was popularly believed that contact with menstrual blood was a prime danger. It was two hundred years before it was clear that men infected women too.

The great pox became somewhat less virulent within a hundred years – a widespread rash the colour of raw ham rather than pustules and death averted for perhaps twenty years. Some must have wished it was not: lumps could appear everywhere from infected lymph glands, as well as bone pain and warts around the anus; penile skin could decay and have to be cut away. The disease, of course, had many levels of severity and, usually, there was a prolonged lull during which the sufferer seemed to have got better. Then rubbery tumours grew in soft tissues and viscera and there could be multiple symptoms including angina caused by bulging of the aorta, blindness, deafness, numbness of the legs leading to paralysis . . . and madness.

The treatment was isolation, semi-starvation, enemas – and mercury: mercury potions to be ingested, mercury sweat tubs, salves of mercury to rub on the skin and to blister the penis. When codpieces strutted through the courts of Europe those of some syphilitics were effectively medicine chests, their penises within wrapped in mercury-treated bandages. Mercury was highly poisonous (and ineffective) and added to a sufferer's woes, particularly in the eighteenth century (when the great pox was renamed syphilis): the dosages became so strong that jaws, tongues and palates were ulcerated, hair and teeth fell out and noses were destroyed – silversmiths made false ones to cover the gaping holes of the nostrils.

Gonorrhoea, popularly the clap – caused by the gonococcus bacterium – was ten times more common than syphilis and its symptoms, while debilitating, far less cruel: a characteristic yellow discharge from the penis, painful urination, swelling and acute tenderness in a testicle; conversations about 'pain in the cods' were heard in every coffee house in Restoration London.

Up to the late fifteenth century, treatment ranged from washing the genitals in vinegar to plunging the penis into a freshly killed chicken. Thereafter, mercury was the popular choice, as it was for the great pox, together with blood-letting, purging, semi-starvation, steam baths and bed rest with as many coverings as could be endured. Luckily the purulent penis cleared up in a month or two, like a runny nose, before the mercury did too much damage. But scars of the infection could create fibrous bands that constricted the urinary passage, in later life necessitating its painful dilation with a curved metal instrument.

For several hundred years venerealogists advised that 'if the Man make long stay in the Woman's Body, and through the excessive Ecstasy, Heat and Satiety, welter and indulge himself in that Coition . . . [it] is much the sooner way to attract the *Venom*, than quickly withdrawing'. Better still was to avoid passionate women. There was nothing so dangerous, men were counselled, as a woman who enjoyed coition – which ultimately led to the schizophrenic Victorian belief, expressed by the physician William Acton, that it was 'a vile aspersion to say that a virtuous woman is capable of sexual feeling'.

Venereal treatments did become more effective during the nineteenth and early twentieth centuries, but it wasn't until the arrival of penicillin during the Second World War that syphilis and gonorrhoea appeared to be conquered – until, that is, the last couple of decades when sexually transmitted diseases began to increase and are proving progressively resistant not just to penicillin but to subsequent antibiotic developments.

In the circumstances one might suppose that in everybody's interests men would use protection as a matter of course; to reverse Madame de Sévigné's complaint about the condom in the seventeenth century, what is available today is only a spider's web against full enjoyment while armour against infection. But many who do use protection, don't all the time, while others simply won't – with those in middle age being the worst offenders, being the least likely to use a condom with a new partner. It is as if penis-possessors have a genetically programmed aversion and regard their member in a condom as something like Munch's *Scream*. They use different terminology but are like their forebears who preferred to risk spending 'one night with Venus, the rest with Mercury' – ending up with their 'pike bravely bent' as Shakespeare has it.

In earlier centuries men could perhaps put up some defence for their behaviour. They were unaware that they spread venereal disease as well as caught it. And before the vulcanisation of rubber, the condom was an unattractive proposition. The first, after the arrival of the great pox (and marketed as a preventative against disease, not a contraceptive barrier), was a tricky linen hood that fitted precariously over the glans and under the foreskin and was held in place by ribbons cinched around the scrotum. It was still going strong in the eighteenth century (ribbons available in regimental colours), continuing to give 'but dull satisfaction', the diarist/biographer James Boswell wrote. Mostly he indulged without, and contracted the clap seventeen times in nine years. Later, the condom was of washable sheep's intestine and extremely unpleasant to don. 'Often my tool, stiff as a boring-iron, would shrink directly the wet gut touched it,' commented the pseudonymous Victorian 'Walter' in his possibly spurious autobiography, *My Secret Life*. He too 'went bareback' and accepted the consequences, time after time 'taking the clap, which laid me up some weeks, and made it again needful to open my piss-pipe by surgical tubes'.

Game, set, mismatch

Men may have grown out of the past's wilder dreads about women's sexuality.[8] But however experienced, however at ease with the dimensions of their penis and their confidence in its cooperation, they enter unknown territory in every sexual encounter, sometimes even with a partner they know well.

Their dilemma is the mismatch between women's sexuality and their own. Inherently, women require stimulation involving their whole body, which takes time; inherently, men just want friction applied to their penis and to move to intercourse as quickly as possible, which is why two thousand years ago Ovid counselled them 'not to sail too fast and leave your mistress behind'. Penis-possessors' interest in foreplay, unless they have developed consideration and restraint, can be limited: 'a quick rub at the clitoris as if to erase it', as one woman has put it, 'and some prodding about for the G-spot – if they have heard of it', before they get down to the business of penetration and ejaculation. For many men, the sexologist Magnus Hirschfeld wrote, 'any other loveplay is a ridiculous nuisance'.

When they do, on average, they're finished five times more quickly than women take to reach the tipping point (see Part 4, 'The Violent Mechanics'); little wonder women often think that men are plain selfish ('The activity, the orgasm was all his, all his' – *Lady Chatterley's Lover*, Lawrence). The actress Lillie Langtry, mistress of Edward VII when he was Prince of Wales, when asked if he was a considerate lover replied, 'No, a straightforward pounder.' More men than not probably are.

Do men lack the emotional intelligence to understand women's sexual needs? The charge has frequently been made (feminism has described the penis as 'the eye that sees everything and understands nothing'). Despite decades of information that the female orgasm is clitoral, not vaginal, men find it difficult to accept the paradox of Darwinian biology that the site of a partner's sexual satisfaction, the clitoris, is divorced from the site of intercourse; and the unpredictability of a woman's orgasm remains a mystery to them. And many continue to at least half believe that it's the penis deep inside a woman's body that should make her swoon. And swoon she should, like the heroine Jordana in Harold Robbins' *The Pirate* (just a typical for-instance) who, when the 'beautiful ten inches' of the gigolo Jacques 'slams into her like a trip hammer', predictably 'Somewhere in the distance [hears] herself screaming as orgasm after orgasm ripped her apart' so that 'Finally she could take it no more. "Stop," she cried. "Please stop."' If a woman doesn't swoon – and according to the voluminous Sex in America survey of 1994 only half of all women ever orgasm, only half of these do so regularly, and 19 per cent never orgasm at all – men tend to think it was because she wasn't 'putting her mind to it' (Kinsey). And they feel resentment – which is why, according to Sex in America, half of all women fake it ('Yes, yes, yes!') out of consideration. Heterosexual men sometimes have a sneaking suspicion, however the sex has turned out, that their partner has got more out of it than they have.

Hardly irrefutable supporting evidence, but the mythological Greek prophet Tiresias, who spent seven years as a woman before regaining his male gender and, supposedly, being in a unique position to know, informed Zeus that when it came to the pleasure of sex, a woman scored nine out of ten to a man's one.

Hurtful as penis-possessors may find it, a 2009 survey found that 29 per cent of women said they get more out of food than sex.[9]

Yet women have greater latent sexuality than men. After a single orgasm men need time out; their penis goes limp and their responses switch off. But women are capable of leaping from orgasm to orgasm, skimming like a stone across water, until they're physically drained, even into the foothills of old age: Kinsey recorded the case of a woman in her sixties who through intercourse and self-masturbation had twenty orgasms in as many minutes.

For decades feminists have asserted the superiority of women's sexuality – and in particular of the clitoris over the penis. The bud-shaped little organ hidden in the upper folds of the labia they have claimed is not only virtually inexhaustible but is 'the only human organ purely for pleasure' – and it carries up to twice as many nerve endings as the penis. Freud, they've argued, had he not thought that 'the sun revolved around the penis' (Erica Jong), might instead have theorised about male envy of the clitoris, not the reverse. That, feminists have pointed out, is in essence what the Aranda tribe of central Australia do by practising ritual subincision, in which the underneath of the penis is sliced open, often along its entire length, the initiate thereby 'menstruating' and thus stealing women's power – 'split penis' in Aboriginal derives from the word for vagina. Adding insult to injury feminists have also stressed that the default setting of human life is female – that every penis in the womb starts as a clitoris before hormones 'sex' the brain of the to-be male and maleness, therefore, is a kind of birth defect. The penis is 'only an elongated clitoris'; and it retains the mark of its female heritage: its dark underskin and the thin ridge or seam, known as the raphe, which runs from scrotum to anus, are remnants of the fusion of the vaginal lips.

Feminists have also delightedly emphasised Kinsey's finding that while males reach their sexual peak between fifteen and seventeen, women are not fully responsive until thirty, by which age men are in decline.

Emphasising the penis's unpredictability and limitations, feminists in the '70s extolled the virtues of the vibrator, with which, Masters and Johnson found, women were able to reach fifty consecutive orgasms. A full-page advert in an English newspaper read: 'It [the vibrator] doesn't stay out with the boys. It's never too tired. And it's always available.' In America feminists gleefully exclaimed, 'No penis goes at 3,000 revs a minute nor is it available with external clitoral stimulator', and suggested that men were an encumbrance to a woman's sexual pleasure with the curiously memorable slogan: 'A woman needs a man like a fish needs a bicycle.' A man might respond that the vibrator is to the penis what the moon is to the sun: similar to look at but lacking heat. But many men are likely to react as did a group at an exhibition of sexual paintings by Betty Dodson, who were 'hostile and competitive' in front of one that showed a woman using such a device, 'one virile stud saying emphatically, "If that was my woman she wouldn't have to use *that thing*"' (*Sex for One*).

Whatever penis-possessors' underlying anxieties about penis size and penile aesthetics, they are exceeded by anxiety about something else: their performance. Yet their capacity for self-deception is greater still. They are like the golfer who remembers those shots which hit the green in one and go down in two, convincing himself that they represent his normal game and choosing to forget the numerous hacks into the rough. Sexually speaking, most men like to believe they play off par; that, like a character in *Lysistrata*, their 'cock is a veritable Heracles invited to dinner'. To which the feminist historian Rosalind Miles has a retort: Phallus in Wonderland.

Spend, spend . . . spent

For centuries men were told that their supply of semen was severely limited and that frequent ejaculation not only drained this supply but damaged their blood vessels, made them weak (and, in China, deprived their brain of nourishment) and could even shorten their life. In the East men practised yoga techniques to enjoy intercourse without emission – an Indian adept, it was said, was able to smoke a pipe during coition without being troubled to climax.

Such a disciplined approach was not for Westerners: they ejaculated every time they got the chance – and only then worried that their 'spends' (an Elizabethan coinage) would put their account in the red.

Intercourse was not the only source of withdrawals, of course: there were involuntary nocturnal emissions; and voluntary masturbation.

Nocturnal emission so worried some Greek and Roman men that they slept with flat lead ingots against their genitals, the 'comparative chilly nature', Pliny the Elder recorded, helping to keep them from arousal to 'venereal passions and the libidinous dreams that cause spontaneous emissions'. Nocturnal emission so worried penis-possessors in the Middle Ages, convinced that if it occurred a succubus (female demon) had had intercourse with them in their sleep, that they placed a sponge dipped in vinegar between their thighs on next going to bed; a safeguard, they believed or hoped, against demonic night-time sexual assault. Less malodorously, Victorian physicians advised men that nocturnal emission could be avoided by 'keeping dreaming thoughts pure'.

The Greeks and Romans thought masturbation a bit unmanly but that it didn't matter much. (The ascetic Greek philosopher Diogenes masturbated in the open air rather than be hostage to 'unkindled desire' and was praised for his reasoning. He recommended masturbation because it was easily available and inexpensive. 'If only,' he wrote, 'one could satisfy one's hunger by rubbing one's stomach.')

Most religions including Hinduism, Islam, Buddhism and Taoism have always been relaxed about masturbation. The Hebrews on the other hand, instructed to go forth and multiply, deemed the activity a crime, even meriting death. Judaism isn't keen on it now. As Shalom Auslander related in *Foreskin's Lament*, when he was young a rabbi told him that 'when I died and went to Heaven I would be boiled alive in giant vats filled with all the semen I had wasted during my life'. Christianity trod much the same early path, even teaching that masturbation was a graver mortal sin than adultery. Some theologians taught that masturbation was possession by the devil.

It would be difficult to say whether the early Church fathers abhorred woman or the penis more. Woman caused the fall from Eden that made man want sinful sex – which was any sex that wasn't intentionally procreative. Until then, it was taught, Adam had known sexual desire but not sexual craving and erected only by will. To Tertullian in the second century, woman was 'the gateway of the devil' and 'a temple built over a sewer'.

But the penis was complicit in carnal desire and curbing its activities in the Middle Ages reached frenzied heights.

Intercourse with a wife was forbidden when she was menstruating, pregnant or nursing; on Fridays (the day Christ died), Saturdays (in honour of the Virgin Mary), Sundays (the Resurrection) and Mondays (in memory of the dead); on feast days and fast days; and during Lent, Advent, Whitsun week and Easter week – which ruled out most of the year. Only one position was permitted, the man on top – and pious men and women were instructed to wear a chemise cagoule, a heavy nightshirt with a hole in the genital area so that bodily contact was as limited as possible. Intercourse was never to take place in daylight or with either partner naked.

The list of penances for transgressions was long. Nocturnal emission, for example, seven days' fasting; masturbation, twenty days. Intercourse with the woman on top incurred a penance of partial fasting for seven years. Intercourse 'not in the proper vessel' (anal) or oral attracted the same penance as murder. Coitus interruptus was punished by two to ten years' penance with the alternative, during the eleventh century, of self-flagellation for monks – who routinely flouted the rules of chastity – or whipping by the parish priest for the laity.

'It is hardly too much to say,' wrote the authoritative G. Rattray Taylor in *Sex in History*, 'that mediaeval Europe came to resemble a vast insane asylum.'

The emphasis on sin diminished by the Enlightenment; the age, however, introduced a new sexual neuroticism. Dusting down the centuries-old belief that a man's semen was finite, it proclaimed masturbation, the most frequent and wasteful means of loss, to be a specific and crippling disease. By Victorian times the disease had a name, spermatorrhoea, and now its causes included all illicit and even excessive sexual activity. Spermatorrhoea was claimed to damage the nervous system, lead to impotence and, in the final stage when a man's ejaculation become uncontrollable and non-orgasmic, to idiocy and death. So obscure was spermatorrhoea's aetiology that everything from tuberculosis to a red nose was diagnosed as a symptom of it.

There was more cause for worry. The medical profession revived the ancient haematic theory – which postulated that semen was extracted from blood in the testicles – and earnestly cautioned men to greater ejaculatory frugality, pointing to the high cost of manufacture; an ounce of semen, it was claimed, was the equivalent of losing two pints of blood. If men needed further prompting, some physicians held that the semen a man retained was reabsorbed into the blood, thereby increasing his vigour; others went further, maintaining that retained semen was vital for the maintenance of secondary masculine characteristics.

How often a man could have ejaculatory intercourse became an issue. A thousand years earlier the Chinese gave detailed instructions. 'In spring [he] can allow himself to emit semen once every three days in summer and in autumn twice a month,' advised the *Principles of Nurturing Life*. 'During winter he should save it and not ejaculate at all. The loss of yang energy by a winter emission is a hundred times greater than spring.' The *Secret Instructions Concerning the Jade Chamber* was rather more liberal. Strongly built men over fifteen could safely ejaculate twice a day; thin ones once a day. Strongly built men of thirty could ejaculate once a day, weaker men once in two days. But at forty a man was to limit himself to once in three days, at sixty once in twenty days and at seventy once a month – 'except the weak ones who should not ejaculate any more'. Victorian physicians were more prescriptive. Most advocated once a week as safe; other voices warned that more than once a month was not, and they included the strident Americans Sylvester Graham and John Harvey Kellogg (both of whom blamed meat-eating for all carnal passions and each of whom created a foodstuff to dampen ardour – Graham the sugared brown biscuit still sold under his name, Kellogg the cereal flake). Some physicians counselled wives to lie still during intercourse so that husbands expended as little semen as possible.[9]

What was misguided mainstream medical belief and what was outright charlatanism is impossible to disentangle. Quacks abounded – treating the guilt-ridden bourgeoisie was highly lucrative. A common swindle was to detect semen in a man's urine under the microscope, indicating 'leakage' and the onset of spermatorrhoea. One reputable physician said that two-thirds of his male patients either had or thought they had the disease.

On the basis that prevention was better than cure, devices were developed to prevent masturbation, denounced as 'self-pollution', and the lesser evil of nocturnal emission. Some were rudimentary: a tin ring with a serrated inside edge that slipped over the penis and caused the wearer pain should night-time erection occur (a quality product in steel with individual spikes also available). Others were more intricate: lockable cages that prevented the wearer from making contact with his genitals or constrained 'longitudinal extension'; galvanic belts of zinc and copper plates, which generated a current if activated by 'secretions of the body'; rubber drawers through which water or cold air was pumped. An ingenious invention was a harness that activated a phonograph on erection, to awaken the wearer with music and save him from himself; should the male be an adolescent, the device could set off an electric alarm in the parents' bedroom.

A few men went so far as to have their foreskin pierced with silk threads that they fastened together on going to bed.

There were a variety of procedures to deal with spermat-orrhoea or masturbation. Physicians embedded potassium and chloral hydrate in men's penises 'to blunt the venereal appetite'; blistered perineums with poison and applied suction cups to draw blood; applied hemlock poultices to the genitals and injected tepid water into rectums; inserted metal, rubber or porcelain 'eggs' into rectums to massage prostates into health. Circumcision was popularly prescribed, much to the satisfaction of extreme moralists on both sides of the Atlantic, including John Harvey Kellogg who was so enthusiastic about the treatment that he advocated it should be done without anaesthetic 'as the brief pain attending the operation will have a salutary effect upon the mind'.

Until the nineteenth century Western culture had no tradition of circumcision. It became fashionable among the aristocracy of continental Europe after King Louis XVI of France was operated on for phimosis (a too-tight foreskin that makes erection agonising, even impossible), which prevented him from intercourse with Marie Antoinette for seven years. Queen Victoria chose to have her sons circumcised, making it de rigueur among the English upper classes.

What made circumcision common among the proliferating nineteenth-century middle classes on both sides of the Atlantic was the hysteria about masturbation; removing the foreskin helped its prevention, doctors declared, and also cured bed-wetting and other conditions. By the outbreak of the First World War such claims had diminished and the medical profession touted circumcision as being 'hygienic' – fathers were not only encouraged to have their newborn sons snipped, but to belatedly enjoy the benefits themselves.[10]

At the height of the panic about spermatorrhoea, many of society's ills were attributed to the moral degeneracy it brought about. Anti-masturbation movements were formed; families were urged to expose adults who habitually indulged in 'the deed of shame'. A boy who did, it was advised, could be identified by 'his shifty glance and the way he pulls his cap down so as to hide his eyes'.

By the end of the nineteenth century spermatorrhoea had lost its hold on the medical profession and the popular imagination. But the obsession with masturbation had a long twilight: men still alive remember when teachers pinned up the trouser pockets of those caught with their hands in them, and the dire warnings that masturbation resulted in hairy palms or even in blindness. As late as the 1930s (when anti-masturbatory devices were still being patented) some of the backstreet museums of anatomy, which had proliferated in Victorian times, were still open. Among the dubious exhibits and wax effigies displaying the Secret Diseases of Men, one cabinet remained dark inside – until someone stood in front of it and a sudden electric light showed the leering face of an idiot and the placard: LOST MANHOOD.

POWER CUTS

No man can say there's never been a time that his penis has disobeyed the command to stand and deliver. For a man normally in full working order such a temporary break in transmission is of no lasting consequence. But for other men, their desire constantly outstrips their capability. Such men are impotent – literally without power. When Henry VIII was intent on offloading Anne Boleyn as his wife and spuriously had her tried for treason and adultery she ridiculed him for being 'without potency' (and, for good measure, that he was 'not skilful in copulating with a woman') – an insult to the royal other head that did nothing to ensure she kept the one on her shoulders.

Impotence comes in degrees. An erection may pump sufficiently to penetrate but then, inexplicably to its possessor, lose interest in proceedings, leaving the recipient of it, as one woman has described, feeling like 'trying to stay afloat on a life raft that is slowly deflating'. An erection may inflate but so weakly that it cannot penetrate, 'pushing at the door', as Fanny

Hill relates of a client, 'but so little in a condition to break in that I question whether he had the power to enter, had I held it ever so open'. For other men, even the semblance of an erection is a distant memory, 'their sex lives in long oblivion/Or if they try, it's hopeless; although they labour all night long at that limp and shrivelled object, limp it remains' (Juvenal).

Limpness has nothing to do with another form of erectile dysfunction: the penis-possessor has no problem with standing and delivering, the problem is that he delivers too quickly – even before he gets as far as penetration in some cases. Medical definitions of premature ejaculation vary: in less than a minute and a half of penetration in one commonly accepted definition, within ten seconds in another, within six thrusts in a third. In his best-selling *The Case of Impotence As Debated In England* (1700), Edmund Curll mocked the condition, writing:

> There are many men whose Penis very readily rises, nay,
> lifts its self up in a most proud and ostentatious Manner;
> but then its Fury is soon spent; like a Fire made of Straw,
> the Moment it approaches its Mistress's Door, it basely falls
> down at the very Threshold, and piteously vomits out its
> frothy Soul . . .

It's been argued somewhat unconvincingly that there's no such thing as premature ejaculation, that men who climax quickly are simply doing what their ancestors did, the theory being that our ancestors once had the sexual characteristics of other primates such as chimpanzees – whose couplings have been timed at four to seven seconds. 'Normal' ejaculation, according to the theory, should therefore be redefined as 'delayed' ejaculation.

Be that as it may, according to the literature perhaps a quarter of men are premature ejaculators. But the numbers of those estimated to suffer impotence are even more astonishing:

half of those over forty experience some degree of it, with five in a hundred totally without function, a figure that rises to around twenty in a hundred in the over-fifties and continues upwards – erections, like teeth, weren't designed to last into old age. An estimated 150 million men in the West are either unable to attain, or to sustain, an erection, ten million of them American, two million of them British.

The agents of erectile dysfunction are complex: psychological, emotional and physical – and in the case of premature ejaculation, a neurological hair trigger. Hippocrates blamed his impotence on having an ugly wife. Some of those who believed in the 'spend' theory (which still had currency in the last century, with Ernest Hemingway among its adherents) blamed women for draining their battery; contradictorily, many impotents throughout history have blamed a partner's lack of sexual enthusiasm, and do so today. Until at least the seventeenth century men thought that witchcraft was the root cause of impotence – if a man's blood was on fire and his penis was not, what else could it be?[11] Actually, seventeenth-century women blamed coffee houses, a pamphlet against the new establishments containing the insult: 'They come from [them] with nothing moist but their snotty Noses, nothing stiffe but their Joints, nothing standing but their Ears.'[12]

In Christian Europe, impotency gave grounds for divorce from the Middle Ages to the seventeenth century (it remains so under Islamic law), indeed it was the only grounds, and it was ascertained by ecclesiastical courts. A man might have his genitals submerged in ice water so that the veins in his scrotum could be examined for constrictions; he might be ordered to a curtained bed with his wife and left for an hour or so, after which the sheets would be examined for signs of emission; or he might face trial by a group of 'honest matrons', as happened in York in 1433 when Alice Scathloe sought a divorce from her husband John. At the top of a house in the city a fire was lit, food and drink brought in, and the women removed most of their clothes, kissed John's face and neck, danced around him, exposed their pudenda and let him feel their breasts. As a last resort they 'warmed their hands by the fire, ticked his testicles and stroked his member' – without result. The court decided in Alice's favour.

In time, impotency as grounds for divorce was taken out of ecclesiastic hands and became part of civil law; and cases were sometimes witnessed by as many as fifteen clerics, physicians, midwives and magistrates.

There was little sympathy for husbands deemed impotent, with no consideration given to the fact that those who contested the charge were likely to wilt under the pressure of having to prove otherwise in public. The examining matrons in John Scathloe's trial cursed him and spat on him; an unnamed husband who appeared before an eighteenth-century civil court in Rheims in France – where the accused went behind a screen to produce ejaculatory evidence of erection, while the witnesses waited around the fire – was ridiculed, as a record of the proceedings shows:

> Many a time did he call out: 'Come! Come now!' But
> always it was a false alarm. The wife laughed and told them:
> 'Do not hurry so, for I know him well.' The experts said
> after that never had they laughed so much nor slept as little
> as on that night.

In relatively recent times received wisdom was that impotence was entirely a psychological problem. Current understanding is that three-quarters of cases, discounting those attributable to the natural deterioration of old age, are due to medical conditions – cardiovascular disease, high blood pressure, diabetes – all of them often undiagnosed. Obesity is another common cause. There is a long list of prescription medications (including antidepressants, diuretics and hypertensives) that cause impotence in many men. As drinking and smoking do in many, many more.

The penis has no head for alcohol; as the sage porter in *Macbeth* remarks, it 'provokes the desire, but it takes away the performance'; beyond that, however, alcohol can lead not just to impotence but to testicular atrophy and penile shrinkage. So does heavy smoking, which also depletes the penis of elasticity (the substance elastin) needed for erection. The attempted erection of heavy drinkers who are also heavy smokers often resembles the balloon left after the Christmas party. The obese frequently suffer shrivelling of the testicles, caused by their fat being converted into the female hormone oestrogen. One in five cases of impotence is due to the penile arteries, which are no thicker than the tines of a fork, clogging up with cholesterol.

Yet activities regarded as healthy are not without dangers. Urologists say that seemingly harmless knocks sustained in youth during exercise are sometimes to blame for a percentage of impotency – 600,000 men in the USA, it's estimated, are impotent because of participation in contact sports. The constant jolting of horseback riding is a factor in some other men's lives (Hippocrates noted that many of the Scythians, who spent virtually all their lives in the saddle, were impotent – though he blamed their custom of wearing breeches) – and cycling in the lives of many, many others. Normal sitting doesn't put weight on the perineum, the area between the scrotum and anus. But cycling does and, the *Journal of Sexual Medicine* reported in 2005, it compresses both an artery supplying the penis with blood and a nerve necessary to sensation. Moderate to total erectile dysfunction can result, with heavy riders coming off worst; additionally, ultrasound tests have shown, as many as half of those who regularly cycle a couple of hours a day develop stone-like calcium deposits in the scrotum.

Anything that causes impotence is also likely to cause infertility – the semeniferous tubes of one in six infertile men have been damaged playing sports, for instance. In heavy smokers and drinkers, sperm numbers plummet and sperm develop abnormalities including reduced motility and loss of sense of direction. All of this said, a quarter of cases of impotence are psychologically rooted: stress, depression, guilt, a worry about not measuring up, the last prevalent among the inexperienced. A doctor can establish whether a problem is psychological, or not, by calling upon a curious fact about male physiology: that three to five times a night during rapid-eye sleep the penis erects and stays erect for between fifteen minutes and an hour. Why this is so is not definitively understood; perhaps to oxygenate a poorly oxygenated part of the body, perhaps, in computer terminology, simply to run a check on the hard drive.

A man has no conscious control over this functioning – and if his problem is in his mind rather than elsewhere, then his penis will boot up normally. In the 1970s, assessing the state of affairs could be cumbersome. Experiencing impotency in early middle age and refusing to believe it had anything to do with his heavy drinking and that his was a case of brewer's droop, the novelist Kingsley Amis took his flagging libido to a psychologist, who provided him with an electric contraption (in his novel *Jake's Thing* he called it a 'nocturnal mensurator') to which futilely he wired himself at night-time in the hope of recording penile activity. Today, a doctor is likely to send a man home with a small snap gauge, usually consisting of three bands of increasing strength that break under different levels of erectile force, if any.

So humiliating is erectile dysfunction that prior to the

arrival of Viagra in 1998, before which most treatments were lengthy, vague or invasive, nine in ten sufferers didn't consult a doctor. Premature ejaculators, who might have been helped with 'mind over matter' techniques, counselling or medication, chose to live with their shortcomings ('Then off he came/& blushed for shame/Soe soone that he had endit' – 'Bishop Percy's Loose Songs', 1650); as did those suffering impotence, no doubt raging at their treacherous best friend's inability to achieve satisfactory elongation. The Earl of Rochester, much given to fornication with prostitutes, poured out that rage in 'The Imperfect Enjoyment', after a rare experience of both impotence and premature ejaculation (the 'thunderbolt below' turning 'dribbling dart of love') with a woman he loved:

> Base recreant to thy prince, thou dar'st not stand.
> Worst part of me, and henceforth hated most,
> Through all the town a common fucking post,
> On whom each whore relieves her tingling cunt
> As hogs on gates do rub themselves and grunt,
> May'st thou to ravenous chancres be a prey,
> Or in consuming weepings waste away;
> May strangury and stone thy days attend;
> May'st thou ne'er piss, who didst refuse to spend
> When all my joys did on false thee depend.

Desperately seeking solutions

Ernest Hemingway was only thirty-eight when impotency started to encroach. It led him to physically attack a fellow writer, Max Eastman, whom he wrongly believed was spreading tales 'and playing into the hands of the gang who are saying it'. Until he died Hemingway was in denial. In his posthumously published 'fictional memoir', *True at First Light*, he wrote about the

old, well-loved, once burnt-up, three-times restocked,
worn-smooth old Winchester model 12-pump gun that
was faster than a snake and was, from 35 years of us being
together, almost as close a friend and companion with
secrets shared and triumphs and disasters not revealed as
the other friend a man has all his life.

Sadly for Hemingway, whose attempted restoratives included
doses of synthesised testosterone, nothing restocked his penis-
rifle; at the end of his life all he could rely on extending was
the metaphor.

There's nothing more despairing for those who still feel
what Eric Gill deemed 'the seethe of tumescence' but who
have nothing with which to answer the call, try as they might,
'as cocks will strike although their spurs be gone' (Earl of
Rochester). Which is why, throughout time, men have sought
ways of stiffening their resolve. Rulers, despots and several
popes have found that taking a young woman into their bed
could work miracles – though not for the aged biblical King
David when he was 'stricken in years' and 'gat no heat'. His
servants provided the fair virgin Abishag, but however hard
she ministered to him, 'the king knew her not'.

Not having access to fair virgins, most of history's possessors
of malfunctioning penises have turned to supposed aphrodisiacs
to beget heat. Eating the genitals of animals, on the apparently
logical assumption that like would help like, is a practice as
old as history in all cultures. What became known as the
'doctrine of signatures' – which stated that something with a
resemblance to something else could be of benefit to it – in
genital terms included types of fish, some at least penis-shaped
and all redolent of the smell and slipperiness of sex, with eels
(naturally) and oysters (the best-known vulvar aphrodisiac)
heading the list. Fruits, vegetables and roots with some male

genital similarity also had their advocates: penile asparagus, celery and carrots, for example, testicular artichokes, truffles, broad beans, tomatoes (the 'love apple' of the sixteenth century), orchis bulbs (once, in fact, called dogstones) and apricots – 'a-prick-hot' a popular Renaissance pun. Garlic, onions, spices and black and chilli peppers were also sought because they quickened the pulse and induced sweating (like sex) – and onions and nutmegs, thought doubly effective because they were testicular as well as hot. The man-shaped roots of ginseng and mandrake, the latter known since biblical times and in the Middle Ages said to grow only where the seed of a hanged man had fallen (a tale that pushed up prices), were highly prized, as were figs (another of the few aphrodisiacs seen as vulvar in appearance, once cut open) and mushrooms, regarded as a symbol of intercourse, the stem thrusting up into the cap like a penis into a vagina.

The East still has faith in traditional sexual stimulants. Asia's once plentiful seahorses are in sharp decline largely because the Chinese so believe in their sexual efficacy (not only fish but persuasively erection-like in swimming upright) that they pulp them in their millions to make an aphrodisiac broth; in Indonesia the cobra is becoming increasingly scarce because concoctions of its penis are more popular than Viagra. And China's conviction that seal penis, tiger penis, bear penis and rhino penis can salvage a faltering erection has contributed to bringing these species to the brink of extinction. The rhino is hunted not just for its genitals but for its horn – the 'doctrine of signatures' at its most logically illogical – as once was the narwhal in the West for its tusk, often sold as unicorn (another tale to push up prices).

History abounds with measures more desperate. Men have swallowed virtually everything imaginable – alabaster, pearls and metals including gold – as hardening compounds.

Assyrian women rubbed a man's penis with oil containing flecks of iron; some Roman men, on the advice of Pliny the Elder, ate excrement; Elizabethans and their French brothers tried urinating through their wife's wedding ring or through the keyhole of the church in which they married; in the early twentieth century hundreds of men (including Freud and the poet/playwright William Butler Yeats) subjected themselves to vasectomies on the advice of an eminent Austrian physiologist who was convinced it would 'reactivate' them (and cure baldness). Hope springs eternal, which is why impotent men in some primitive tribes in the Amazon still ask their comrades to blow on their penis, in the way one blows on a fire's dying embers.

From the eighteenth century, the medical profession in the West took the view that toning up the man would tone up his below-par penis. In consequence men undertook regimens of fresh air and exercise, tonics and beef-steak, cold baths and scrubbing with rough brushes. Penises were cleaned out like a blocked chimney with surgical instruments; corrosive chemicals were deposited at the prostatic end of the urethra. 'Electro-therapy', heavily advertised in the newspapers, was popularly prescribed; by the early twentieth century it was fashionable for well-to-do men with erectile problems to wear a battery-pack belt that passed a current to their genitals, which purported 'to improve sexual vigour by massage'. The first suction and vacuum pumps became available to force blood into the erectile tissue, kept there by a constriction ring.

With the rise in the standing of surgeons in the nineteenth century, the idea that animal genitals could be the key to impotence made a comeback, with a twist. The doyen of European physiologists, the seventy-two-year-old Charles-Édouard Brown-Séquard, injected himself with a composite of dog and guinea pig testicles and caused a sensation by claiming that doing so had enabled him to 'visit' his young wife every day without fail. As the century turned, others, the serious and the charlatan, went the whole way to xenotransplantation – in America, John Brinkley used the testicles of Toggenburg goats; in France, the Russian Serge Voronoff used grafts from chimpanzees and baboons. In fact, Voronoff had begun by transplanting human testicles – into millionaires – but demand exceeded his limited supply of executed prisoners and he had to find another source.[13]

Thousands of men around the world in the 1920s went under the knife for the supposed benefits of what were known as 'monkey glands' and by the 1930s there was triumph in the air, with monkey glands being claimed a success by both those carrying out the procedures and their recipients. Sadly, the whole affair turned out to be a massive example of the placebo effect at work – and just as Brown-Séquard succumbed to a cerebral haemorrhage soon after his astonishing announcement (his young wife having run off with a younger man), xenotrans-plantion died a death.

Throughout recorded history men have helped half-hearted erections with a variety of external supports; in the 1950s surgical techniques were developed to put the supports inside the penis. Bone and cartilage weren't successful, but twenty years on silicone was: a pair of pliable rods could be implanted in the spongy columns of the penis – with the drawback that it was permanently extended, needing to be bent down when dressed, bent up for sex, and a man was always aware of its presence. Inflatable models with some mechanical parts have overcome the problem, but more men opt for the complex alternative, an implant with a separate fluid reservoir located in the belly, attached to a squeezable ball inserted in the groin or in the scrotum like a third testicle.[14]

It was in the 1980s that a reliable non-invasive treatment for impotence emerged – by accident. French surgeon Ronald

Virag inadvertently injected papaverine (an opium alkaloid for treating visceral, heart and brain spasm) into an artery leading to his anaesthetised patient's penis instead of the artery he intended, and was taken aback when his patient's penis sprang to attention. Virag was rather upstaged by the British urologist Giles Brindley, who was intentionally investigating penile artery dilation as a treatment for impotence. He did more than report to a convention of fellow urologists in Las Vegas the success that he'd achieved with the beta blocker phenoxybenzamine (a treatment for hypertension): he showed them. Having injected himself before taking the platform, he walked among his audience, erection in hand, to prove that no implant was involved. Soon others were promoting dilatory compounds for self-injection (yes, a small prick for a bigger one).

And then came Viagra (which had nothing to do with the near-anagrammatic Ronald Virag), which like papaverine and phenoxybenzamine relaxes the smooth-muscle cells of the penile blood vessels – but with the enormous advantage of only having to be swallowed. Viagra, the first approved oral treatment for erectile dysfunction, coined from 'virility' and, hyperbolically, 'Niagara', became the fastest-selling drug in history. (Among the first customers were hard-working males in the porn industry, the drug enabling them to 'maintain wood' without the help of 'fluffers', young women whose job was to lick them into shape just before the director called 'Action!') And it seemed that without pain, inconvenience or prolonged treatment impotent men could at last with certainty hold their head(s) up high . . .

A PRICE TO PAY

A man may be distraught when he can't get an erection — and terrified when he can't get rid of one.

A prolonged and agonising erection — priapism — happens when the blood in the spongy cylinders of the penis that engorge during arousal doesn't return to the circulation, which normally happens after orgasm. Four to six hours later the blood trapped in the penis has the consistency of thick oil — and if a doctor doesn't remedy the situation by sticking in a needle to let the blood out, there will be damage to blood vessels and nerves that might make getting an erection impossible. If left untreated for twenty-four hours, gangrene and even amputation may follow.

Priapism can be a side effect of certain medical conditions or of recreational drug use. But almost any effective impotency treatment can cause the predicament, including over-pumping a penis or leaving on a restricting ring for too long, both of which can also result in permanent impairment. And priapism may not be the only, or the worst, outcome of attempting to encourage, strengthen or prolong erection.

The overwhelming majority of aphrodisiacs are ineffective and harmless; some are dangerous. Most of these work by irritating the mucous membrane of the urinary tract and genitals to assist the necessary blood flow; a very few are psychotropic, which is to say they work on the mind to induce sexual desire – yohimbine, made from the bark of a West African tree, is the best known. And all are powerful poisons, including such innocent-sounding plants as crowfoot (a member of the buttercup family), periwinkle and henbane, as are mandrake, toad venom and Spanish fly, made from crushed southern European blister beetles. The mimicking of the natural sexual response comes with various side effects (in the case of yohimbine, panic attacks and hallucinations) but sometimes with greater costs, including gastrointestinal bleeding, renal failure, lung or heart damage – and even death. So deadly are some aphrodisiacs that they are illegal in most parts of the world, but they, or compounds containing them, are frequently smuggled or, like everything else, found on the Internet. The deadliest aphrodisiac of them all, made from the testes of the blow- or pufferfish, is not illegal in China and Korea, to which its use is largely confined. A gland in the fish contains tetrodotoxin, which is a hundred times deadlier than cyanide – and the merest trace left after its removal means certain death. Some three hundred men succumb every year.

It's unlikely that anyone has lost his life directly from a penile implant, though lurid and unlikely stories of inflatable devices being activated in public by mobile phones or bleepers, or exploding and causing haemorrhage and death, regularly appear on the Internet. What is true is that nearly three-quarters of those who've had a device fitted have not been happy with the results – for them, bio-hydraulic sex really isn't like the real thing. In the early days, inflatable models were prone to malfunction, leak or break; and in the parlance of several lawsuits, device components were said to 'migrate'. Implants have become more reliable (though they remain liable to cause infection); Viagra, however, has made them a minority choice, usually for those made impotent by prostate cancer surgery and beyond chemical conjuring.

And yet Viagra and its similar competitors have not proved to be the Holy Grail for all men in search of a viable erection (for a start, two in ten men find no such thing results). The tablets should only be used on medical advice − and not by those being treated for hypertension, high cholesterol, liver or kidney problems, diabetes or obesity. Most at risk are those with heart disease on drugs containing nitrates: the potency tablets work by releasing nitric oxide in the penis and the interaction of related compounds can cause a catastrophic drop in blood pressure. Men without medical conditions commonly experience varying degrees of dizziness, nasal congestion or nausea, or more unpleasantly, temporary visual disturbance. But for those with the identified conditions the price can be sudden hearing loss or blindness, respiratory failure, stroke or heart attack. Or death. There are no definitive statistics, but in the dozen years since Viagra came to the market, hundreds of the twenty-five million men worldwide who have popped pills for potency have popped off in the process.

As many as a quarter of all men bypass their doctor to obtain their supply elsewhere − the Internet again, usually. Some who do are unaware that they have an underlying medical problem and are therefore ignorant of their risk. The majority know the state of their health, but either can't bear to expose their problem to scrutiny to obtain a prescription or just think, what the hell − such is the driving force of the biological imperative to have sex.[15]

If the force be with them in middle or old age, even men in apparently good health face the danger of a heart attack or stroke in the act of committing adultery: excitement and exertion (and sometimes the stress of infidelity) can lead to the fatal rupture of an aneurysm. It's happened through history to the eminent as well as the anonymous. The British prime minister Lord Palmerston died (1865, at eighty-one) while having sex with a parlourmaid on a billiard table; French president Félix Faure died (1899, at fifty-eight) in a brothel having sex with his secretary; American vice president Nelson Rockefeller died (1979, at seventy) having sex with a mistress in her apartment. Staff at Japanese love hotels, where not-young businessmen traditionally take young women, are unsurprised to open a room and find one occupant gone, the other still there but checked out in a way he hadn't anticipated: a final flaring of the lamp, they say in Japan. The French talk romantically of coital death as *la mort d'amour*. The world for centuries has colloquially described it as death in the saddle.

PART THREE NOTES

1. A Florentine mob might have emasculated Michelangelo's Boy David had they got the chance.

    In 1504, thirty years after the work was finished, a mob hostile to the 'new paganism' of the Renaissance stoned the statue, which had to be guarded for five months until a modesty girdle of twenty-eight copper leaves was attached; the girdle remained in place until 1545.

    In 1857 a plaster cast of the eighteen-foot-high David was presented to Queen Victoria, who forthwith donated it to the South Kensington Museum, now the Victoria and Albert. The museum, believing anecdotal evidence that the queen had been shocked by David's 'insistent nudity', took the decision to have a proportionally sized fig leaf of stone made and kept in readiness for any royal visit.

    Such prudery might seem to belong to another age. Yet in 1986 the V&A had the covering on standby in anticipation of a visit by Diana, Princess of Wales.

2. Jews in Hellenic cultures, given to public nakedness in the baths and gymnasium, were often persecuted for being circumcised, which led many, particularly during the reign of the Roman emperor Hadrian, to attach a stretching device called the *Pondus Judaeus*. Even in periods when Jews were not being persecuted, some men abandoned their faith and got to work with the *Pondus Judaeus* to improve their social and economic standing.

    The uncircumcised have from time to time discriminated against the circumcised – and vice versa. In the ancient Muslim world, Africans were circumcised rather than castrated when sold as slaves and any who tried to stretch their remaining foreskin with weights were sometimes put to death. Outbreaks of violence today have erupted in South Africa between the circumcised Xhosa and the uncircumcised Zulu and in Kenya between the circumcised Kikuyu and the uncircumcised Luo. After a disputed election in 2008, roaming gangs of Kikuyu cut off the foreskin of any Luo they apprehended.

    During the Nazi era many Jews underwent surgical 'uncircumcising'.

    Skin-graft surgery was once an option but is rarely carried out now because of the different colour and texture of the graft against the existing penile skin. But foreskin restoration using weights is still popular among some Jews and non-Jews alike. Restoration can take several years, depending on how much skin was left after circumcision.

3. The world's first penis transplant was carried out in a Chinese military hospital in 2005, on a forty-four-year-old man whose own organ had been reduced to a centimetre-long stump in an accident. A newly brain-dead twenty-three-year-old was the donor. In a fifteen-hour operation surgeons attached arteries, veins, nerves and the corpora spongiosum, the lagging around the penile piping of the uretha. The penis largely regained its functions.

4. Did Shylock in *The Merchant of Venice* want to castrate Antonio? It's an interpretation some scholars have made of the Jew's terms for lending money to Bassanio that if Antonio failed to honour his surety he would forfeit 'an equal pound of your fair flesh to be cut off and taken in what part of your body pleaseth me'.

    By the later trial scene the bond has changed, in Portia's words, to 'a pound of flesh to be by him cut off nearest the merchant's heart'. But the interpretation of Shylock's initial stipulation can be argued. Elizabethan playwrights including Shakespeare often used 'flesh' for penis, and a penis, according to those who postulate the reading, weighs about a pound; that seems an over-generous estimation, though they invariably add, 'more or less'.

5. The hijras of South Asia are castrated males who dress as women and see themselves as a third sex, neither male nor female. Some are transsexual; others have become hijras simply as a way of surviving; living in communities, on the margins of society, they recruit boys who have been rejected by or fled from their families. Hijras demand money in the streets, sing and dance at weddings and offer sexual services.

6. Male to female surgery has become relatively straightforward. The scrotal sac is cut open and the testicles removed from it. The penis is then cut open, its contents put to one side and the 'shell' turned inside out and pushed into the body cavity to form a vagina. The scrotum is then fashioned into a labia and a piece of the erectile tissue from the penis contents used to create a clitoris. The procedure can be so expert it can't be detected by the trained eye.

Female to male surgery is more expensive and less successful – a fully operable penis isn't achievable. A scrotum is formed from the patient's labia and synthetic testicles inserted. The clitoris is elongated, with the help of skin grafts, to become a penis, which can be used for urination but not ejaculation. Some opt for an internal prosthetic device so that the penis can be pumped erect allowing intercourse to be simulated.

7. The legend of a sharp-toothed demon who hid inside a young woman to castrate young men – and who was defeated by the local blacksmith's ingenuity in fashioning a metal penis to break the demon's teeth – is celebrated in the Shinto Kanamara Matsuri (Festival of the Steel Penis) in Kawasaki, one of the diminishing number of similar Japanese penis festivals.

8. A woman knows she is the mother of a newborn child. Generally speaking, a man has to take her word for it that he is the father.

Some men have always had reason to worry whether they were a child's biological father or not. From the 1920s various blood testing methods became available, though they were difficult to perform and often inconclusive. In the 1970s, however, a test of HLA cell proteins was developed, with a 90 per cent probability of establishing paternity. DNA testing – 99.99 per cent or higher certainty – came in the 1980s.

A worldwide survey reported in the *Journal of Epidemiology and Community Health* in 2005 found that one in twenty-five fathers was unknowingly bringing up another man's child.

In Elizabethan England, men had additional anxieties. A man welcomed the birth of twins if they were two girls, or a boy and a girl, but was less pleased if both infants were male: the belief was that one was born from the right testicle and the other from the less virile left, meaning that one son (which?) would be lacking in manly ways. But triplets were a matter of greater concern – triplets meant that another man had had congress with his wife. Curiously, a medical belief in sixteenth-century Italy was that a child could be conceived from the semen of up to seven men, each contributing a portion of his character.

The Hottentots in southern Africa once practised hemicastration – the removal of one testicle – to prevent the birth of twins, which were regarded as bad luck.

9. Did the earth move? Women of the Kagaba tribe in Colombia hope not. Tribal belief is that if a woman moves during intercourse the earth will slip off the shoulders of the four giants holding it above the water.

10. Europe, like Japan, generally didn't buy into the hygiene argument. In Britain circumcision remained popular among the better-off, but rates shot up under the new National Health Service set up after the Second World War; they dropped dramatically once parents had to pay for it – the 50 per cent rate of 1950 is almost zero today. America's enthusiasm for circumcision has only waned in the last decades, from 95 per cent to an overall 60 per cent, though it is half that in the eastern states.

   Circumcision is a human ritual going back to prehistory. It's central to the religion of Jews (who traditionally circumcise at eight days) and Muslims (at puberty), but also to the beliefs of many tribes in Africa and Australian Aboriginals. Among some primitive peoples males are considered neuter until the foreskin, which is seen as being feminine because it bears some resemblance to the labia, is removed. The Greeks and Romans were appalled by circumcision when they encountered it among the Egyptians and Israelites. Esteeming the tapered, fleshy, nipple-like portion of the foreskin as the defining feature of the male, they even passed laws forbidding such 'mutilation'.

   There has been considerable debate as to the sexual pros and cons of circumcision to the adult male. Some specialists consider it beneficial in that it delays orgasm. Others maintain that it robs a man of nerve-rich erogenous tissue and sensation in the permanently uncapped glans, blunted over time to become about 'as sensitive as a kneecap'. Others have dubbed circumcision 'penile rape'. The probability is that it doesn't make much difference either way.

   The foreskin is much larger than might be thought: up to 15 square inches.

11. In Renaissance France the belief was that a man with a grudge against a new bridegroom could make him impotent by calling out his name while at the same time breaking off the point of a knife in the marriage chamber door.

   Similar beliefs occur throughout time – Athenian males worried that black magic could be used to harm their penile potency.

Enemies or sexual rivals often inscribed a lead tablet with a curse aimed at another's generative organ and buried it in the grave of a boy – the ghosts of the 'premature dead' were believed to wander the earth and were prepared to wreak evil until their natural lifespan was fulfilled.

12. Insult the penis, insult the man – and mankind has been doing that with the phallic gesture throughout history.

Two thousand years ago the Romans were giving each other what antiquarians called the phallic hand, just like the Ancient Greeks: the fist clenched, the tip of the thumb thrust between the first and second fingers like the tip of a penis. The *mano fico* (the 'fig') in Latin but fuck you in any language except Japanese – the ever-so polite Japanese don't have such a gesture, or any swear words either. Probably older is the *digitus impudicus* (impudent finger): the same meaning, with greater economy. The 'fig' is still popular in Mediterranean countries, particularly Italy, but the 'finger' is almost universal, though with some variants: the first finger instead of the more usual middle one, or the two fingers held up closed together or spread apart. Not so in Iran and some other Middle Eastern countries where the equivalent of the 'finger' is what in the West is the cheery gesture of encouragement: the thumbs up. Which can lead to serious misunderstanding either way.

A newer variation is the forearm jerk, in which one hand is slapped down on the bicep of the opposite arm. Another, popularised by the comedian Jasper Carrott, air-sketches a protuberance from the forehead with forefinger and thumb: dickhead – the same meaning indicated by pressing the back of the closed hand against the forehead, with the first finger and thumb extended. A gesture common in some African and Caribbean countries in which the five digits are extended with the palm forward, meaning you have five fathers (you bastard), has been adapted and given more emphasis by holding the hand backwards against the forehead: not just a dickhead, but a dickhead five times over.

In recent years women have favoured crooking the little finger and waggling it: little dick and droopy with it. With much the same psychology behind the Israeli poster campaign of the previous decade, in 2007 the Australian government produced a series of TV commercials in which women gave the (little) finger to male drivers driving too fast. When a woman in Sydney emulated the commercials the enraged recipient of her signal smashed a bottle on her car, and pleaded not guilty in court on what he considered the justifiable grounds that his manhood had been impugned.

During the Second World War, the British Political Warfare Executive, responsible for black propaganda, impugned Adolf Hitler's manhood in a startlingly explicit way. The British people cheered themselves up during the war by singing, to the tune of 'Colonel Bogey', that Hitler had only got one ball (true: as medical examination revealed in young manhood, the right testicle of the foetal Fuhrer had failed to make the normal journey from the developing abdomen and down the inguinal canal into the scrotum). But the PWE devised a far greater insult. It took a photograph of Hitler standing on a balcony, put a penis purportedly his, in his hand – a very small penis and circumcised, to feed the rumour that he was a self-hating Jew – and produced a postcard from it captioned 'What we have, that we firmly hold', quoted from a speech he made in Munich in 1942. Some two and a half thousand copies were dropped over Germany in March 1944 before the operation was cancelled, a government minister saying he would rather lose the war than win it with the help of psychological pornography.

An intriguing question: is the 22-foot erection of the Cerne Giant, the chalk figure carved into a hill above the English town of Cerne Abbas, a phallic insult or not?

For most of the last three centuries the 180-foot Giant was thought to be a fertility symbol of the Romano-British period, or possibly Phoenician, Celtic or Saxon; and it was the custom during these centuries not only to mount a maypole on the site during spring planting and summer harvest, but for infertile women, newlyweds and couples about to be married to visit it at other times, and perhaps sleep on the Giant's penis for marital luck. But in the late twentieth century scholars noted that the earliest written reference to the Giant was made only in 1694 (as a three-shilling payment, found in the churchwarden's accounts, for tidying up the figure); and that in 1774 the Reverend John Hutchins in his guide to Dorset had stated that the Giant was 'a modern thing', which had been cut in the previous century by the landowner, Denzil Holles, an MP who opposed Cromwell with great hostility and got thrown into prison by him several times. On this basis, and because the Giant carries a (phallic) club, scholarly speculation became that Holles's motive in having the Giant cut was to give form to the epithet by which Cromwell was mocked by his enemies: the English Hercules.

The probability is that the Cerne Giant *is* an ancient fertility figure, and an extraordinary one at that. If it is not, it's the biggest phallic insult ever made.

13. In San Quentin, prison doctor Leo Stanley had no such problem – he not only had access to the testicles of the executed but of those who died naturally and whose bodies went unclaimed; he carried out more than 600 transplants with volunteers from among the inmates.

14. In the 1980s surgeons attempted to rewire penile arteries, rather like carrying out miniature heart bypasses, with a high incidence of failure. Now, in China, experiments are being conducted in which arm muscle is being grafted into dysfunctional penises to encourage lift-off; in America drug-releasing stents are being fitted into penile arteries in much the same way that larger versions are fitted into heart attack victims, for the same reason: to give the blood flow clear passage. How these procedures turn out remains to be seen.

15. There have been reported deaths of young men who used potency pills for increased sensation in their lovemaking. In 2009 a twenty-eight-year-old Muscovite bet two women four thousand dollars that he could handle a twelve-hour sex marathon with them. He took multiple Viagra and won his bet – but died of a heart attack. Perhaps the most extraordinary case of death-by-sex without stimulants was recorded in the 1753 Paris medical journal *Recueil periodique d'observations de médecine et de chirurgie* of a young man who killed himself by having intercourse eighteen times in ten hours.

PART FOUR

CAMERA!
LIGHTS!
ACTION!

Love is a matter of chemistry, but sex is a matter of physics.

Anon

THE PRICK OF THE BRAIN

That blood engorges the penis to make it erect seems too obvious to mention. But this was not always what was believed. The Ancient Greeks thought that air rushed from the liver to the heart and down the arteries to inflate the penis in much the same way as we pump up a bicycle tyre. The medieval Church wanted to believe that the spirit of God elevated the organ for procreational purposes but could hardly convince itself: too many erections quite evidently arose when procreation was the last thing on the penis-possessor's mind.

In the late fifteenth century, Leonardo da Vinci recorded what really made the penis stand on end, after attending the dissection of a hanged man and subsequently dissecting other hanged men himself. He wrote in his diary that he had seen 'dead men who have the member erected . . . all of them having great density and hardness and being quite filled by a large quantity of blood'.

Why a hanged man dies with an erection was not properly understood until four hundred years later, when the role of the nervous system in normal erectile functioning became clear –

and from which it could be deduced that the sudden fracture of a hanged man's cervical vertebrae produced a violent stimulus of the nerve centres, springing open the penile blood vessels (it can also happen in other types of violent death). The first definitive description in Western medical literature of the part blood plays in erection was published by French physician Ambroise Paré (unaware that da Vinci had got there before him seventy or eighty years earlier). Whether in the years between da Vinci and Paré the rumbustious Rabelais came to know that blood, not air, raised the penis, he certainly knew that a hanged man's gave a final salute. 'Isn't it a jolly death,' he wrote cheerfully (*Gargantua and Pantagruel*), 'to die with a stiff john-thomas?'

Primate penises vary from species to species: that of man's closest relative, the chimpanzee, for instance, is conical, rather like the cardboard horns blown at parties and New Year, whereas man's is cylindrical. And man is alone among the primates in producing his erection entirely through blood pressure: other primates, like most mammals, have a penile bone – a baculum or os penis – which is deliberately flexed when an erection is required; far quicker, and more reliable, than human hydraulics. Evolutionary biology deduces that man's early ancestors must have been so equipped: humanoid females, like most other primate females, would have copulated with numerous partners in succession and speed for the male was of the essence because of the likelihood of being hustled aside by the next in line. Perhaps many impotent men now would lament, as did (the potent) Henry Miller (*Tropic of Cancer*), that 'the bony structure is lost in man'. But why was it lost? An unusual interpretation of Genesis suggests an explanation: that it was from Adam's penis bone, not his rib, that God created Eve. The basis for the speculation is that men and women have the same number of ribs – men aren't missing one – and, anyway, biblical Hebrew had no specific word for rib: the word used, *tsela*, means any supporting strut. Evolutionary biologist Richard Dawkins theorises that the loss of the penis bone was down to female selection: a bone-dependent erection said nothing about the penis-possessor's health, whereas an erection dependent on blood pressure alone spoke volumes: poor rigidity was a warning that a male might be genetically weak and unlikely to father healthy offspring.[*]

Life, as penis-possessors become frustratingly aware, would be so much simpler if getting an erection was a matter of will, as it was for Jean Cocteau, poet, painter, playwright, novelist, designer and cineaste, whose party piece as a young man was another manifestation of his versatility: lying naked on floor, table or chaise longue, he would bring himself to full ejaculatory orgasm, without touching his penis, by the power of imagination alone – to applause all round. There are few Cocteaus able to override the parasympathetic wiring into which erection is hooked, just like breathing and digestion. Erection is involuntary: it is not under the penis-possessor's command, but a reflexive response to multiple psychogenic and sensory stimuli. The best erections, of course, travel both the physical and psychological pathways.

Things happen when such stimuli zing down the spinal column. Local nerves release nitric oxide to make the penile arteries open and blood floods into the thousands of tiny, sinusoid veins fanning out from the arteries, filling the penis's central corpora spongiosum (through which the urethra runs) and the corpora cavernosa, the twin, sponge-like chambers either side of it. Like a flower in slow-motion photography, the penis swells, lengthening, thickening from base to tip, elevating. And as the blood presses against the penis's casing

---

(tunica), 'lock-in' is created, the penile head, where most feeling is centred, glistening slightly from the blood beneath – it isn't skin but membrane, similar to that of the inner eyelids and lips but remarkably thinner.

It takes only two ounces of blood to fully charge an erection, which is eight to ten times the amount of a penis's normal supply. In a young man the process takes seconds, in a man of fifty at least twice as long, and in those of advanced years, time can only be calculated in patience – but with physical encouragement the ageing penis, like an old warhorse, can respond to the trumpet.

The pressure of blood in the fully erect penis is at least twice that in the main circulation; cross-cultural research, however, indicates racial differences in the erections that result. Generally speaking black penises are less hard than white penises, which in turn are less hard than Asian penises. And, generally speaking, when penis-possessors are in a standing position, the erections of blacks hover at the horizontal, those of whites a little above it – though one in five reaches 45 degrees – and those of most Asians rise vertically, more or less tight against the belly. There can be a correlation between rigidity and elevation, but not necessarily. The erection of a very, very few men is so powerful that they can carry heavy objects suspended from it – Armand in Jean Genet's homosexual novel *The Thief's Journal* lifts a heavy man on the end of his. Such penis-possessors are often eager to demonstrate their prowess: national servicemen in the army during the 1950s tell tales of individuals who could support the billet's galvanised metal fire bucket, even in some cases with a pair of army boots inside it; at Princess Margaret's house parties on the island of Mustique in the 1960s former gangster John Bindon is alleged to have balanced a full half-pint glass or suspended five empty ones by their handles. Sexually speaking, of course, the optimum erectile state is that which allows satisfactory penetration.

The human male doesn't even wait to be born to have erections: he has them, as ultrasound scans show, in the womb. So many newborn males greet the world with an erection that the sexologist William Masters in his earlier obstetric days set himself the challenge of trying to cut the umbilical cord before it happened.

Mothers become hot and bothered because their infant son has erections and take his hand away from his penis, yet sometimes they stroke it to soothe or to lull him to sleep; in some cultures, women suck the infant penis – 'the open secret of every ayah, "the native treasure" on whom the British memsahib relied so heavily for the care of her children in the days of the Raj' (Jonathan Gathorne-Hardy, *The Rise and Fall of the British Nanny*). And, sometimes, ambivalently, mothers play with the infant penis, as did the nurses of Rabelais's Gargantua who, 'when he began to exercise his codpiece', rubbed it between their hands 'like a roll of pastry, and then burst out laughing when it raised its ears'. When he was a small boy, the nurses of the French king Louis XIII similarly amused themselves, making him so proud he showed his governess, saying: 'My cock is like a drawbridge. See how it goes up and down.' He tried to show his father but could not pull off the trick. 'There is no bone in it now,' he said mournfully. 'But there is sometimes.'

The adolescent male is prey to frequent and indiscriminate erection. Sights, sounds, smells, anything or seemingly nothing at all can do it, including innocent activities such as sliding down the banisters or riding a bicycle; in his autobiography, William Butler Yeats wrote that at the age of fifteen, after going for a swim, he covered his body with sand and 'Presently the weight of the sand began to affect the organ of sex, though at first I did not know what the strange, growing sensation was. It was only at the orgasm that I knew ... It was many days before I discovered how to renew that wonderful sensation.' The motion of a bus or train is often the inciter, as is acute self-consciousness. 'It seems that I can't go up to the blackboard in school, or try to get off a bus, without its jumping up and saying "Hi! Look at me!" to everybody in sight,' confesses Philip Roth's Portnoy.

The experience can be embarrassing – and humiliating if spontaneous ejaculation follows.

The majority of males experience their first 'spend' by masturbating; most of the rest find it happens involuntarily while they sleep and dream; but a few are unfortunate enough to be caught out in public. Even many of those who think themselves sexually aware find that the sudden sensation of loss from the body, and the violent pounding of their heart, fills them with anxiety that something is wrong with them. As he recounts in his autobiography *Flannelled Fool*, man of letters T.C. Worsley had the daylights scared out of him by his housemaster at Marlborough who remarked: 'You might find some white matter extruding from your private parts, Worsley. Don't worry about it. It's only a sort of *disease*, like measles.'

In adulthood spontaneous ejaculation is rare but it can happen: it happened to that practised philanderer James Boswell while he was playing kneesy with a woman at the opera. Erections caused non-sexually become rare too (there are exceptions: the party piece of the poet Rupert Brooke was to dive into the River Cam and emerge with his penis at the perpendicular – it impressed Virginia Woolf no end when they went skinny-dipping), but they can spring a surprise at any time. As the middle-aged writer in Hanif Kureishi's *Intimacy* is aware at the yoga class he attends with his wife, the sight of attractive women in bright leotards reflected in the long wall mirrors 'taking up adventurous positions', makes his penis press against his shorts 'as if to say "Don't forget that always I am here too!"'

Such a wilful erection merely flickers; others more wilful still can blatantly firm, especially if physical contact is made, unintentionally and unavoidably as can happen in a crowded lift or the crush of an underground train. Some penis-possessors seek such contact, deliberately rubbing up against women in public places, as indulged in by Samuel Pepys and others before him and since. Frottage, as the act is termed, and known with coarse good humour by the Victorians as 'bustle punching', is today a preoccupation of Japanese salarymen in Tokyo, where every year over two thousand arrests are made for various forms of molestation.

Dance halls legitimise a kind of frottage, when couples dance close together. If flickering or firming occurs there is likely to be discomfort on at least one side of the occurrence, or possibly both. Or not. At seventeen the poet Sylvia Plath wrote in her diary about a partner who 'On the dance floor held me close to him, the hard line of his penis taut against my stomach,' adding, 'And it was like warm wine flooding through me.'

Deep inside the brain is the pineal gland, little bigger than a grain of rice, which the medical profession in the seventeenth century thought was the junction between mind and body (the philosopher René Descartes believed it to be the seat of the soul). They also thought it had some sexual function, hence the name; medical texts of the period refer to it as 'the yarde or prick of the brain'. In fact, sexual activity does not arise here but in the hypothalamus in the core of the brainstem, which coordinates basic drives including sleeping and waking patterns – utilising the hormone melatonin produced in the pineal gland – which, in turn, are largely governed by how day and night are perceived by the eyes. In a sense the seventeenth century was not wrong in wondering if the pineal gland was a sex accessory, though it would have been nearer the mark to say that about men's eyes. Men 'fuck with their eyes', the Spanish say (more politely, *mirada fuerte* – 'strong gazing'). 'In Andalusia the eye is akin to a sexual organ,' wrote David Gilmore in *Aggression and Community: Paradoxes of Andalusian Culture*. The film critic of the *Guardian* newspaper David Thomson once suggested that women were not auteurs because they lacked the primacy of the male gaze, which is essentially voyeuristic; men on the other hand, he suggested, feel with their eyes – just like the camera. Thomson's wife put it succinctly in a paraphrased Andalusian way: men 'see with their pricks'.

And they see temptation everywhere:

> Each day the penis is prey to sexual sights in the streets, in stores, offices, on advertising billboards and television commercials – there is the leering look of a blond model squeezing cream out of a tube; the nipples imprinted against the silk blouse of a travel-agency receptionist, the bevy of buttocks in tight jeans ascending a department store's escalator; the perfumes aroma emanating from the cosmetic counter: musk made from the genitals of one animal to arouse another. (*Thy Neighbor's Wife*, Gay Talese)

Indeed men (and their complicit addendum) sexually scrutinise almost all women regardless of their attractiveness to them: legs and buttocks ahead of them, breasts, groin and legs coming towards them. It is a largely subliminal activity; men are like an anti-virus program, monitoring, monitoring. Women generally do not behave in this way. A man here or there will catch their attention but not the passing panoply of body parts. Women have a much lesser level of visual stimulation – which is why most are not aroused by male genitals and, Kinsey found, fewer than one in five wants the light on when they make love. Almost all men do at some point in their lives.

Various studies have tried to establish how often the penis-possessor thinks about sex, with wildly different results: once every seven seconds (Kinsey), for example; at least once every twenty-four hours (*International Journal of Impotence Research*) – the latter study claiming that British men think about sex more frequently than any other nation in Europe. What is certain is that sexual thoughts flicker in the background of a man's visual cortex all day and almost all night. When they do, the sexual-pursuit area of his hypothalamus, an area nearly three times larger in his brain than in a woman's, lights up like a slot machine and, neuro-imaging shows, hot spots of blood-flow erupt. Constantly, intermittently, he thinks about sex: idly, innocently or less innocently, fantasising.

> I would give all I possess
> (Money keys wallet personal effects and articles
> of dress)
> To stick my tool
> Up the prettiest girl in Warwick King's
> High School

wrote the poet Philip Larkin in a letter to his friend Kingsley Amis, the kind of fleeting fancy not unfamiliar to most men, albeit not in verse.

The penis-possessor's brain is his most sexually active organ, only off duty when he is asleep, and not always then, any more than his penis.

Testosterone, the male hormone, manufactured in specialist cells (Leydig cells) that lie between the sperm-producing coils in the testicles, is the root cause. In the womb, testosterone shapes both the male genitals and brain – where it shrinks the communication centre, reducing the hearing cortex to make room for the part that processes sex. At puberty testosterone surges through the male body, increasing ten to thirty-five times, deepening the voice, creating body hair – and making desire overwhelming. The difference in the levels of desire between penis-possessors and non-penis-possessors, says the evolutionary psychologist David Buss, is staggering: 'like the difference between how far the average man and woman can throw a rock'.

As he wrote in old age (*The Summer of a Dormouse*), the playwright/novelist John Mortimer at Oxford shared Bustyn, his scout (servant), with the future Archbishop of Canterbury Lord Runcie, who one day asked him why Mortimer always had young women in his room and why he wore purple corduroy trousers; to which Bustyn enigmatically replied: 'Mr Mortimer, sir, has an irrepressible member.'

Sexually and metaphorically speaking, all men wear purple corduroys. Their desire – and the urge for variety that goes with it – is irrepressible. And, sometimes at least, 'it leads them astray, causes them to beg favours at night from women whose names they prefer to forget in the morning' (Talese). And to cheat on someone they love. Indeed desire can sometimes be so all-consuming it makes men irrational and believe 'there are some fucks for which a person would have their partner and children drown in a freezing sea' (Kureishi). When men are deprived of women entirely, the scabrous American comedian Lenny Bruce said, they 'will fuck mud'. After visiting Egypt the historian Herodotus wrote that 'When the wives of high-ranking men die, the husbands do not deliver them right away to be embalmed, nor do they immediately hand over very beautiful or famous women but wait till the third or fourth day after death. They do this so that the embalmers may not have sexual intercourse with these women.'

In the (possibly) autobiographic *My Secret Life*, the Victorian Walter put the male craving for intercourse in terms that no penis-possessor would dispute:

> If you can't afford to pay for cunt, or don't know a cunt
> which will take you in for love, your prick is a restless
> article which will insist on the buttocks pushing it
> somewhere or somehow, till the stiffness is take out of it.

Most men frustrated for the lack of a woman, the *Kama Sutra* observes, 'seize the lion'. Some, however, 'must satisfy themselves with the vaginas of other species, mares, she-goats, bitches, sheep . . . or with other men'.

Although bestiality has almost always been universally condemned, before the nineteenth century same-sex male intercourse was widely regarded in Western culture as exhibiting nothing more than a lustful urge; indeed it was de rigueur for an Elizabethan fop to have an 'ingle', a beardless boyfriend poised on the brink of manhood, and to consort with the odd 'Ganymede' or boy-whore (the youth Ganymede was wooed by Jupiter, father of the gods), without being labelled a sodomite. The nineteenth century, however, demonised such sex, labelling it deviant (in the latter half of the twentieth century sex therapists began to prefer the word 'variant' as purely descriptive and non-judgemental).

About half the world's cultures condemn homosexuality, two-thirds of the remainder condone it, and the rest ignore it. Kinsey claimed that human sexuality is a spectrum, and, too, that homosexuality is in the genes, dismissed by many as simplistic determinism. What is true is that more men who consider themselves heterosexual than most people would imagine have had some homosexual contact.

Kinsey divided the ways a man can achieve orgasm (or 'total outlets' as he preferred to say) into six categories: nocturnal emission, masturbation, heterosexual petting, heterosexual intercourse, homosexual activity and animal contact. It isn't likely that more than a few have experience in all categories, unlike one of Kinsey's subjects, the amazingly active Mr King, who kept records of his sexual activity with girls (two hundred), boys (six hundred), countless adults of both sexes including his grandmother, father and fifteen other relatives, and not a few beasts of the field. At the age of sixty-three he demonstrated that he was able to masturbate to ejaculation in ten seconds.

Men who are heterosexual in orientation, and their penises, display a multiplicity of other kinds of variant behaviour.

Whereas for almost all men erotic thoughts about a woman, never mind fleshly contact, are often enough to galvanise their body's gang of hormones and neurotransmitters which lead to erection, for some that is not enough. In his novel *Justine*, the Marquis de Sade wrote about the Comte de X who found it impossible to obtain an erection except by cheating while gambling. The early sexologist Havelock Ellis (bizarrely himself impotent until the age of sixty when he discovered he became aroused watching a woman urinating) was told by a young prostitute about a client who could only reach orgasm when she wrung a pigeon's neck in front of him. Alfred Kinsey interviewed a Congregationalist minister who got erections when he saw a female amputee on crutches.

According to Kinsey the brains of almost all men are intensely curious sexually in a way that the brains of almost all women are not. Men want to experiment. Both de Sade and Henry Miller tried intercourse with cored apples filled with cream. Shades of the film *American Pie* – rock wild man Alice Cooper used to masturbate into his sister's jelly doughnuts. Just as the Ancient Greek Clisyphus 'violated the statue of a goddess in the Temple of Samos after having placed a piece of meat in a certain part', Portnoy in *Portnoy's Complaint* had sex with a piece of liver ('I fucked my own family's dinner!'). A few young men have finished up in A&E after attempting relations with a vacuum cleaner.

TWO TO TANGO

Solitary masturbation (Shakespeare's 'jerks of invention'), is, of course, the bread and butter of outlets, possibly the sole sexual activity of youth but a safety valve for almost all adult males, however healthy their sex life, if only sporadically. (Numerous writers including Pepys, Voltaire, Kierkegaard, Gogol, Rousseau, Flaubert and Walt Whitman have claimed that it made their creative juices flow.) Like so many of the words of sex, masturbation is a nineteenth-century coinage. Before that, certainly from the sixteenth century, 'frig' (from the Latin *fricare*, to rub) was the common term in English. 'Wank' (1940s, origin unknown) is the most popular synonym to the British, replacing 'toss off' (eighteenth century, from the original meaning of finishing a piece of work quickly); Americans largely adhere to 'jerk off'.

Freud maintained that masturbation was exclusively male and was infantile. Kinsey pointed out that half of all women masturbate too, if far less frequently, and when they do are more adept, because they have a romantic imagination. Visual creatures that they are, men want concrete images (which is why they are drawn to pornography) or a residue of one on the retina – starved of female company when living under canvas in the Canal Zone during the 1950s, national service squaddies, according to one, 'would get a fix on one of the handful of WRACS [who were only interested in the officer class], get a good image, and rush off to their tents to wank' (*The Call-Up*, Hickman). Fantasy made flesh: as the playwright Arthur Miller relates in his autobiography *Timebends*, he was once in a bookshop with his wife Marilyn Monroe and watched a man masturbating in his trousers as he watched her.

Being masturbated by a woman, which somewhat dismissively Kinsey categorised only under heterosexual petting, is a considerably heightened experience from going it alone – with the proviso, as Alex Comfort pointed out in *The Joy of Sex*, that the woman 'has the divine gift of lechery', does not treat the penis like a gearstick, and goes about her task 'subtly, unhurriedly and mercilessly'. No one has suggested that this constitutes 'having sex'. But what of oral (orogenital or buccal) intercourse, which Kinsey also included in the same category? In 2010 the Kinsey Institute found that almost a third of eighteen- to ninety-six-year-old Americans thought that oral sex did not qualify – the basis of Bill Clinton's denial of congress with White House intern Monica Lewinsky, having enjoyed her fellatory attentions in the Oval Office. (That Clinton the while smoked a cigar, the most recognisable of modern phallic symbols, added ribaldry to the affair.)

As a source of sexual gratification, oral sex runs vaginal intercourse pretty close for most men. Bill Clinton when American president told an air hostess that it was his 'most favourite thing'; in fact, if it were a case of either vaginal or oral sex only, one man in five says he would opt for oral sex.

The ancient world extolled fellatio as much as bathing.

Indeed, Mesopotamia had the same word for semen and fresh water – both of which fertilised life.

Several ancient cultures, principally India and China, ritualised fellatio (*fellare*, Latin, to suck, first recorded in English in 1887). The Greeks and Romans had something of a hang-up about it. Theoretically they considered fellatio to be unclean but practised it anyway, paying lip service to their higher principles. A variation was *irrumatio*, in which the mouth of another was used as a passive orifice as a penalty or a humiliation – an act that was about power, not sexual satisfaction. For almost all Romans, cunnilingus was unthinkable: the male mouth, the springwell of oratory, was not to be debased.

Despite the strictures of the Church, medieval Europe did not give up fellatio. After the plague closed the bathhouses, however, and during the following centuries that came to believe bathing opened the pores and allowed disease to enter, oral sex may have been a less pleasant experience, but this is unlikely to have been much of a deterrent, any more than the criminalisation of it (for centuries it was deemed to be a type of sodomy). As mini-directories of the thirty thousand prostitutes who worked in eighteenth-century Paris show, many claimed oral sex was their speciality – this despite the century's medical profession considering the act a sign of insanity. Curiously, the libidinous James Boswell, ever complaining that the girls he picked up on the streets of eighteenth-century London gave him the clap, never once, from the evidence of his diaries, asked for the safer alternative.

In the 1940s Kinsey found that only four married men in ten had had oral sex. A quarter of a century later, the Playboy Foundation found that almost two-thirds of all men had been beneficiaries, and nine out of ten under the age of twenty-five. Since Linda Lovelace's performance in the film *Deep Throat*, fellation has come to be seen as a cultural phenomenon in the West. In recent decades the incidence of it has grown exponentially among the young – no longer something to which a couple, who probably had already had intercourse, graduated, but just another element of foreplay.

Oral sex, of course, is a two-way street ('to eat another is sacred,' John Updike wrote in *Couples*). And while there is at least twice as much fellation as cunnilingus among those who are unmarried, largely because of anatomical accessibility in sometimes awkward surroundings, there is equal give and take among those in stable relationships. If the penis is capable of penis envy, it should be of the penis-possessor's tongue.

The reasons why a man finds skilful oral sex unspeakably pleasurable is because the lips of a woman's mouth are infinitely more versatile than the lips of her vagina; and her pliant tongue in the words of Gerstman, Pizzo and Seldes (*What Men Want*) is 'the Swiss Army knife of sex'; and the mouth is further accessorised by the teeth, which can be tantalisingly employed to nibble and nip. The Japanese call fellatio 'mouth music'; indeed a woman who knows her stuff can play a man like a flute.[2] Does fellatio put the man or the woman in physical or psychological control? The question concerns several of the 'ological' disciplines. A man doesn't care; he never quite believes he's so lucky that a woman will take his penis in her mouth. Only a relatively small percentage of men want to ejaculate in a partner's mouth, but the experience is heightened for those who do: there is no interruption necessitated by moving to the coital site as the moment of inevitability approaches – and they are observers of their own climax.

Most women, whatever the depth of their feeling for their partner, however much they might enjoy giving oral sex – if hardly to the extent of the already introduced Jordana in *The Pirate*, who wants 'to swallow [Jacques] alive, to choke herself to death on that giant, beautiful tool' – can't bring themselves to permit buccal insemination. Contrary to the delight shown by women in Internet pornography, many have feelings of disgust; the violence of the final orgasmic thrusts can be distressing; and, even penis-possessors admit, the smell and taste of semen are not wholesome.

The smell of what the eighteenth-century physician John Hunter called 'a mawkish kind of substance' has been likened to seaweed, musk, pollen, the flowers of the Spanish chestnut tree, a greenhouse in summer – all of which may sound alluring – but most penis- and non-penis-possessors agree that semen smells like nothing so much as household bleach. As to its taste, the *Brihat Samhita*, the ancient Sanskrit astrological treatise, suggests it can be like honey; most women would suggest fish, perhaps, or worse: overripe Brie, dirty socks, nasal mucus are descriptions given in modern surveys. In Rick Moody's *Purple America*, Jane Ingersoll muses that 'men's curds' taste 'like toothpaste with a soy sauce chaser'.

Different foods affect all bodily secretions and that applies to ejaculate, though to what degree is debatable. Red meat and dairy produce are said to result in the least pleasant flavour, with asparagus, garlic and onion not far behind. Beer and smoking have a deleterious effect. On the other hand most vegetables, peppermint, parsley, cinnamon and citrus fruits are said to make ejaculate more palatable. In America a powdered drink is marketed made of pineapple, banana, strawberry, broccoli and celery 'at nine times their normal concentrations, together with three essential spices and a select blend of vitamins and minerals', supposedly making semen sweet 'in only twenty-four hours'.

What is indisputable is that, if ingested, the average ejaculate won't make the recipient fat – it contains only one to seven calories.

Location, location, location

How many positions are there for a man and a woman to have intercourse? The great ancient literatures of India, China, Japan and Arabia were obsessed with calculating the permutations, some attempts running into the hundreds.

Classical Greece pragmatically reasoned there were about a dozen. They didn't bother to describe them all (one they called 'the lion on the cheese-grater' still has scholars unsure). But basic physiology and the elimination of the improbably gymnastic – such as the *Kama Sutra*'s 'fixing the nail', in which the prone woman stretches out one leg while placing the other on top of her head – make this difficult to argue with: man on top or underneath; woman on top or underneath, face to face with her partner or reversed; side by side, face to face or with the woman reversed; one or other kneeling or sitting; the woman on hands and knees, her partner behind her; both standing up or only the man – whether or not using the furniture as props, seemingly a fixation among the Chinese.

There was only one position as far as the medieval Church in Europe was concerned; anything other than the man above was perverse. The Elizabethans later advocated the position principally because they believed that face-to-face sex distinguished human sex from bestiality (they would have been horrified to know that pygmy chimps [bonobos], orang-utans and occasionally gorillas enjoy a bit of ventro-ventral activity). There is no woman-on-top sex in Shakespeare but that does not mean there was no woman-on-top sex in Shakespeare's England.

According to ancient sources, interpretations of ancient art and anthropological studies, what is now universally called the missionary position was often adopted in the first great civilisations only when conception was desired, medical conviction being that it ensured 'the proper flow of semen'. But for pleasure it was not that high on the list for the Sumerians,

Indians, Persians, Romans and Greeks, who all favoured woman-on-top sex – the Greek courtesan/prostitute *hetaera* charged the most for the 'racehorse', in which she sat astride a prone client. Whether the Greeks really did have a predilection for heterosexual as well as homosexual anal sex is now disputed, though from the woman's point of view the act obviously had the advantage of protecting against pregnancy, and prostitutes up and down the social scale certainly offered a price; what is certain is that men had a fondness for standing rear-entry vaginal sex, usually for 'quickies' in the street, the woman arching herself against the penetratee (cheapest) or resting her hand on her knees or feet (more expensive). Coincidentally, standing rear-entry vaginal sex was not peculiarly Greek: an anthropological study published sixty years ago identified eight primitive peoples around the world who practised it at the time, 'confined to brief and sudden encounters in the woods'. Traditionally, coital preference in the vast islanded region of the Pacific, as well as in parts of Africa (notably Ethiopia), had different preferences yet again: the most popular position involved the woman recumbent with the man squatting between her thighs. In a variation, the man sat in the lotus position, the woman, also in the lotus position, facing him while squatting on top of his thighs.

Islanders found the European way both indecent and amusing. Kinsey, misreading the journal of a 1920s anthropologist who'd lived among the Trobrianders, in 1948 wrote that Christian missionaries had instructed the natives that only intercourse with the female supine beneath the male was allowable. In fact, missionaries had done nothing of the sort. What had happened was that the natives had parodied the Europeans, joking that the evangelisers must have forced them to adopt the ridiculous position. Kinsey's error is neither here nor there – except that it gave the position, previously in the modern world known as 'male superior' or 'matrimonial', a new demotic designation.

Today, the missionary position is the most common throughout the world, from West to East. Its popularity has been ascribed to a man's psychological need to feel dominant and a woman's to feel submissive; to face-to-face (and heart-to-heart) sex seemingly being the most intimate. From a physiological viewpoint it would appear to be the most natural way for male and female bodies to connect. In *Purple America*, Jane Ingersoll muses that 'missionary style is boring as oatmeal'. But it need not be, with imagination – and using it does not preclude adopting others for variety. Some people never try another way. Many are more adventurous – 'Three-quarters of love,' Casanova wrote, 'is curiosity.'[3]

The violent mechanics

Feminists, including the writer and academic Germaine Greer, objected to the word 'fuck' on the grounds that its original meaning was 'to strike', which therefore made sexual intercourse an act of violence against women. Later, fed up with the habitual use of 'fuck' in almost all contexts, Greer suggested the reintroduction of 'swive', an alternative with a longer etymological history (and its original meaning was non-violent, 'to revolve').

But sexual intercourse requires physical vigour – or violence, hardly a semantic difference – on the part of a man to reach completion; not for nothing did the Ancient Greeks label intercourse 'the violent mechanics'. Which is why, from the moment of penetration, a penis-possessor on average thrusts his way to ejaculation in about 4 minutes – the length of time that has been accepted by generations of sexologists.

But a 2005 survey in Britain, America, Spain, Holland and Turkey, reported in the *Journal of Sexual Medicine*, found the average to be 5.4 minutes. The finding, however, was reliant on self-timing, which might be as reliable as penile self-measurement (couples were provided with stopwatches – the British claimed the longest, 7.6 minutes; the Turks the shortest, 3.4). According to another survey, taken among American sex therapists who drew on their male and female clients' responses, 7 to 13 minutes was 'desirable', 3 to 7 was 'adequate' and 1 to 2 was 'too short' – but not unknown in the lazy Sunday morning leg-over or the snatched 'quickie'. Again in a Rick Moody novel, *Ice Storm*, a man has intercourse with his friend's wife in the front of his friend's Cadillac in 'less time than it takes to defrost a windshield'.

Coitus and copulation, the modern standard terms for sexual intercourse, are hardly in everyday use. But throughout history people have almost always preferred slang expressions, most of them vulgar – and many with a 'violent' connotation.

There have been periods when the acceptable standard words were, in fact, the most common. The earliest and least-known now was sard, first recorded by the Anglo-Saxons, with currency up to the seventeenth century, and which co-existed with swive, the most popular colloquialism for almost six hundred years from Chaucer to the late Victorians ('Do not bathe on a full stomach,' advised a popular self-help book of 1896, 'nor swive'). Another word found in both formal and daily contexts was jape (thirteenth century), which faded as the modern meaning of 'practical joke' established itself. 'Occupy' is curious in that it came into being in the fourteenth century with the respectable modern meaning of being 'in possession of', was a vulgarity for three hundred years, and then became respectable once more. Over two hundred slang terms for sexual intercourse are recorded in English, Old, Middle and Modern. Many have come and gone and exist only as dictionary archaisms: for instance, plough (which goes back to the Greeks and Romans); root and the much older rootle; and foin (from

Old French for fish spear; in fencing, to thrust).

But many old terms are still with us, including the sixteenth-century shag (Shakespeare favoured the variant shog), grind (but in Elizabethan times 'to do a grind'), knock (today usually followed by off or up) – and fuck (which Shakespeare never used). Hump was fashionable in the seventeenth century; roger and bang (prostitutes were bang-tails) in the eighteenth; poke, shaft and screw (contraction of screw driver) in the nineteenth. 'To lie with', used by the King James Bible and Shakespeare, has disappeared, though the confusion between the verbs lie and lay gives rise to America's favourite euphemism, laid.

The twentieth century's contributions to the lexicon include bonk and boff, which like fuck meant, and in other contexts still mean, to hit or strike. Fuck, however, besides being the most frequent expletive – and doing service as virtually any word in a sentence – remains the commonest term for intercourse. Sometimes the variant 'frig' (which is also slang for masturbate) is used. The young Norman Mailer was persuaded to change 'fuck' to 'fug' for the publication of his first novel, *The Naked and the Dead*, in 1948. 'So you're the darling boy', exclaimed the actress Tallulah Bankhead on meeting him, 'who can't spell fuck.'[4]

Eighteenth-century Indian harlots scoffed at the way European males 'scurried' to ejaculation and mocked them as 'dunghill cocks'; they were used to somewhat better, at least with educated higher-caste clients. As Hindu, Buddhist and other erotic literatures make clear, a woman's pleasure should be central to sexual activity and a man should learn to hold back from climax so that she can have as many orgasms as she wishes. Through the sexually orientated spiritual meditation techniques taught for millennia by tantric and taoist masters, a man can copulate without climaxing for a considerable time, even almost indefinitely. And he achieves this by first understanding that orgasm and ejaculation are not the same thing. Ejaculation occurs in the penis, orgasm in the brain — which, of course, triggers ejaculation. Kinsey pointed this out; and that half of five-year-old boys have orgasms which is long before ejaculatory age. Masters and Johnson later discovered that in some men ejaculation doesn't occur until some seconds after orgasm, which makes it incontrovertible that they are separate functions, though for the majority they're simultaneous.

Emperors in ancient China had good reason to learn iron self-control. They were required to keep 121 wives, a precise number thought to have magical properties, and to make love to ten of them every night (their sex secretary kept records), which would have been impossible had they climaxed on every coupling. In more recent times King Ibn-Saud of Saudia Arabia, the first Saudi king, practised the same control — he slept with three different women every night from the age of eleven until his death in 1953. Another was Prince Aly Khan, the international playboy son of the head of the Ismaili Muslims, who in the 1940s and '50s had more than a thousand affairs in Europe and America and was reputed often to make love to a woman in his car as he was being driven between the flats of two others. Aly allowed himself to climax no more than twice a week for fear of debilitating himself.

In recent decades some European men have claimed to have learnt to become multi-orgasmic, to have several orgasms with the one erection and even to have aspired to true sexual ecstasy — an orgasmic state in which, it's said, the orgasm flows through not just the genitals but the whole body, even the skin. Tantric sex devotees (notably pop singer Sting and his wife Trudie Styler) are said to be capable of making love for as long as five continuous hours, which most penis- and non-penis-possessors alike might find a tad excessive.

There are times when every penis-possessor wants to prolong his activity. Most usually put a brake on proceedings by stopping thrusting, asking their partner to stay still or by thinking of something else, the more mundane the better; or they take the temperature down a notch or two by withdrawing before starting over. More riskily some drive to the crisis point and then attempt to short-circuit their responses by squeezing the base of their penis, or pressing hard on the acupuncture point of the perineum midway between rectum and scrotum, or tugging their testicles to the bottom of the scrotum – the testicles rise up during the climactic process. All of these actions, known for centuries, may help to some degree. A few go so far as to don multiple condoms or a condom treated with a mild anaesthetic to dull sensation – which might be to defeat the object of the exercise. Walt Disney apparently sometimes packed his scrotum with ice to prolong lovemaking with his wife.

But whether a penis-possessor is a good lover who ensures his partner's satisfaction, or not; or times his lovemaking so that infrequent happening, the mutual orgasm, can happen, or not; or for his own pleasure or hers or both he has held back – the moment comes when he cannot.

A myriad response has been set off in his body. His diastolic blood pressure, normally as low as 65, rises to around 160, systolic from 120 to around 250. His pulse rate, normally 70 to 80 beats a minute, reaches anything from 150 to 250. His breathing is harsh – a shortage of oxygen. His sense of smell and taste diminish, his hearing becomes impaired, his sense of vision narrows, so much so he may not be able to see objects on either side of him. His scrotum tightens, his testicles, swollen by vaso-congestion – often half their normal size again but in some men as much as double – elevate, in many pulling tight against the penis shaft, in a very, very few even disappearing into the abdominal cavity.

His thrusting becomes shorter, quicker, more furious. The first of his sex accessory organs, the Cowper's glands, come into play. These pea-sized organs, immediately in front of the prostate, secrete a few droplets of an alkaline mucus to neutralise the urethra of any traces of urine, which is acidic and could damage the sperm that are about to follow this route. The droplets form at the urethral opening and may carry a few sperm, which can cause pregnancy – those who practice withdrawal before ejaculation may not be safe. (Not without irony was the secretion once called the 'distillate of love', as, not without irony, coitus interruptus in our time has been termed 'Vatican roulette'.) Meanwhile, the prostate and the winglike seminal vesicles attached to it are pumping a milky protein-rich fluid into the firing chamber at the root of the urethra, a suspension medium to carry the sperm that the vas deferens ducts are simultaneously delivering from the comma-shaped epididymis on the top of each testicle, where they've matured. The sphincter between the prostate and the bladder clamps down, akin to a train track switching points, so that the semen doesn't discharge into the bladder.

The prostate spasms.[5]

Spinal nerves quiver.

Contractions ripple along the urethra.

And with the last thrusts, semen propels from the penis in three to eight spurts, 'the sweetest sensation of a man's life' (*What Men Want*).

The force with which semen exits the penis depends particularly on how powerfully the prostate spasms. Clinical monitoring of men during masturbation has shown that in most the semen merely exudes or spurts an inch or two. In some, however, particularly and unsurprisingly among the young, it can travel two or three feet; Kinsey recorded rare instances of adult males whose ejaculate shot six to eight feet.

Some penis-possessors equate a copious ejaculate with masculinity – just as they might equate a large penis – and consequently have an exaggerated idea about the volume of their own; they may even erroneously believe that the greater the volume the greater a woman's pleasure. A Cruikshank cartoon depicting Sir William Hamilton, his wife Emma and her lover Horatio Nelson catches both notions neatly. As Sir William tries vainly to light a very small pipe and Nelson puffs vigorously on a pipe that is both phallic and reaches to the ground, Emma remarks: 'Pho, the old man's pipe is always out, but yours burns with full vigour.' To which Britain's naval hero replies: 'Yes, I'll give such a smoke I'll pour a whole broadside into you.'

The 1970s pop group 10cc named themselves after what they thought was the average amount of ejaculate. In fact it's 2 to 5cc – less than a spoonful.

Post-coitally, for a moment or for minutes, mildly or intensely, the bodies of the partners spasm. A woman's orgasm may be a passing ripple or a thunderous tsunami outstripping a man's, which is fairly constant, whatever the quality or intensity of the sex. At its most extreme, men are more likely to flail their limbs, groan or shout as if 'suffering the extremes of torture' (*An Analysis of Human Sexual Response*, Ruth and Edward Brecher) – they have very likely put in more effort, however brief, and they have greater muscle mass from which tension must be released. But some women go into the same violent convulsion, rolling their eyes, pounding, punching or kicking their partner, oblivious to pain themselves, flinging themselves feet or yards; a very few intensely reactive individuals lose consciousness for seconds or even much longer – small wonder that the French dub orgasm *la petite mort* (the little death).

So, as the return to normalcy after climax throws the physiological changes into reverse, the neuromuscular tensions abate, pulse and blood pressure subside, blood returns to the circulation and the penis shrinks, simply subsiding or retreating in a series of little hops, like someone crawling backwards, carefully – how was it for him, really? Or her?[6] Only he or she can answer this and their answers might be different concerning the same occasion, whether the sex is perfunctory or prolonged, routine or rampant, rough or tender of a combination of some or all of these. Words are probably inadequate to capture the convolutions of sex. What words can say is that sex is likely to be at its best for one or the other or both when lust and love are in sync, the brain anaesthetised, the body saturated with feeling, warm skin in contact, limbs entangled, the universe in a lover's eyes – a state in which, as Alex Comfort described it, 'while the penis is emphatically his, it also belongs to both of them'.[7]

Save for the very young male who may maintain an erection for several minutes after orgasm and a very few penis-possessors, of any age, who may maintain full rigidity for up to half an hour and who, if sexual activity resumes, may achieve another orgasm, or several, without ejaculation, virtually all men enter a refractory period after sex during which they can't respond to sexual stimuli of any kind. And during it, for a while, their penis is so sensitive that further stimulation of it is unpleasant and even painful. Nineteenth-century marriage manuals recommended that a man who reached orgasm before his wife should continue coital movements until she had been satisfied, but for virtually all men that is physically impossible. And here is one of the greatest mismatches between the sexes: women not only can carry on orgasming if stimulated to it, they don't have a clear-cut non-reactive phase: their descent from the heights follows a fairly gentle curve – and they want to cuddle and talk. Men, on the other hand, whose exertions can sometimes be comparable to heavy labour or the effort of an athlete at full stretch, have fallen off the cliff.

'I think men talk to women so they can sleep with them and women sleep with men so they can talk to them,' the novelist Jay McInerney once observed. Of course men can try to be accommodating, to kiss and caress, to murmur sweet nothings;

and sometimes they do, they do. It isn't that they don't have feelings. But unless they have a pressing reason for getting up, they have an overwhelming desire to go to sleep. They can't help it: the hormone prolactin, released during ejaculation, strongly urges them to sleep so that energy-producing glycogen, depleted by intercourse, can be restored to their muscles. And, too, the more pleased a man is with his performance (his body is flooded with the feel-good neurotransmitter dopamine), the more likely he is to drift away.

On benefits

The penis erect not only gives sexual gratification but, as numerous studies show, can contribute to the health and well-being of both giver and receiver. Various hormones and other chemicals released before and during orgasm help to lower blood pressure, decrease bad cholesterol, improve circulation, mediate pain, and ward off stress – one study suggests that intercourse can be as much as ten times more effective than Valium. Lovemaking can also help to repair tissue, promote bone growth and burn off calories (an average of 85–150 in 30 minutes of activity), as well as dampen food cravings by increasing an amphetamine that regulates appetite. And having sex can even improve brain power – intense intercourse encourages brain cells to grow new dendrites, the filaments attached to nerve cells that allow neurons to communicate with each other; there is some evidence that older people who are sexually active are less likely to have dementia.

Women gain additional benefits: intercourse helps to keep their skin elastic, meaning fewer wrinkles, stabilises their menstrual cycles and reduces hot flushes during the menopause – women who have sex age more slowly than women who don't. But men gain additional benefits of their own. Regular sexual activity reduces the risk of prostate cancer and, in older men, the likelihood of developing benign enlargement of the prostate. And regular activity is an aid to their longevity: a study which tracked the mortality of about a thousand men over a decade concluded that those who had sex twice a week had half the risk of a fatal heart attack compared with those who had sex once a month.

THE 'PRECIOUS SUBSTANCE' REVISITED[8]

Scrotal skin is thinner than the skin anywhere else on the body – indeed it's translucent against a light shone on it from one side in the dark. A woman contributor to *FHM* magazine wrote that she liked to crawl under the duvet with a torch to watch 'how the skin gently shifts and crawls, forming and reforming in mesmerising goose-pimply patterns'.

The cerebellum of the brain is 'corrugated', to increase its surface area and allow more cognitive RAM; the skin of the scrotal sac has similar corrugations (which in *Fanny Hill* Cleland dubiously described as 'the only wrinkles that are known to please') but with a different purpose: to assist heat loss and keep sperm at three degrees below body temperature. The corrugations almost double the scrotal sac's surface – Rabelais poked fun at this when Panurge meets the noble Valentine Viadiere rubbing 'his ballocks, spread out upon a table after the manner of a Spanish cloak'.

Inside the scrotum, the testicles produce spermatozoa at the astonishing rate of seventy million a day. They are the only human cells designed to travel outside the body and constitute only 1–5 per cent of ejaculate, the rest being the fluids from the prostate and seminal vesicles that give them the energy for their journey. To the touch, the testicles seem to be solid lumps, but these two hard glands are like the inside of golf balls, comprising a mass of tiny tubes where sperm are manufactured, a process that takes between two and three months. If unravelled and laid end to end the tubes inside a testicle would stretch over a quarter of a mile.

Sperm are continuously being shuttled from the testicles to the epididymis where they mature, gain motility and the biochemical properties to fertilise an egg, and are then held in a staging area awaiting orders. If not ejaculated they undergo autolysis: they dissolve and are reabsorbed into the body – so much for Baden Powell in the 1920s telling young males who masturbated that 'You are throwing away the seed that has been handed down to you as a trust, instead of keeping it and ripening it for bringing a son to you later.'

Lifespan of a sperm: a month in the staging area, two days inside a woman's body, perhaps two minutes on the sheets.

A healthy sperm consists of a head, a mid-piece, which is its powerhouse, and a tail. Inside the head, which is paddle-shaped, oval in outline but flat, is its package of DNA. The head wears a kind of cap containing enzymes to melt the membrane surrounding the female egg.

On ejaculation the oldest sperm are first out, but the youngest at the back – arriving in the later spurts – beat them to the cervical mucus. If a woman is not ovulating, the juices in her vagina, cervix, uterus and fallopian tubes are acidic, and acid kills sperm. But for the short period when she is ovulating, the normally thick juices clear and become alkaline, giving sperm the green light.

In the last fifty years sperm counts have more than halved (from about 200 million per ejaculate to about ninety million) and are decreasing by 1–2 per cent a year. Scientists have identified numerous possible causes including chemicals that mimic the action of the female hormone oestrogen and are found in plastics and paints, the linings of food cans and disposable nappies, and pesticides. Synthetic oestrogen is a constituent of many drugs, the basis of the contraceptive pill, and finds its way into the water supply. Heat is also implicated; the sperm of men who spend long hours sitting in motor vehicles decrease in number and vigour – a situation considerably worsened for those with heated car seats. The heat from laptops balanced on knees also poses a threat.

But in 2009 evolutionary biologist Oren Hasson of Tel Aviv University entered the debate, saying that stressful lifestyles and pollution can't explain the levels of plunging male fertility. He suggested that 'polysperm' are to blame – men are now producing super-sperm so vigorous that they race past the defences set up by a woman's body. When a sperm penetrates an egg and their chromosomes fuse, all other sperm are supposed to be sealed out. Hasson suggests that super-sperm can be so powerful that more than one can break in, effectively destroying the egg – some small comfort, perhaps, to men who are failing to become fathers to be told that 'their boys' are over-virile.

It's likely to surprise most men that even for the healthiest not all their sperm are the sleek athletes they think them to be: up to 50 per cent of sperm have morphological defects or poor (or no) motility. Only about a quarter of sperm in an ejaculate swim forcefully enough – an average rate of 1.5 millimetres per minute, a speed comparable to a human swimmer in relative terms – to reach the target. In the 1990s evolutionary biologist Robin Baker caused considerable controversy when he claimed that only 1 per cent of sperm have any chance of being an 'egg-getter' – because he believed that sperm consist mostly of two other types, which weren't designed for fertilisation: killer sperm that are on guard to attack the sperm of another man if necessary, and blocker sperm that knot their tails together to form barriers to any such arrivals. Again according to Baker, non-egg-getting sperm change their roles as they age. Most are killers when young and blockers when old. Killers need to be full of energy and movement; blockers need only enough energy to swim out of the seminal pool and travel a little way into the cervix.

The scientific community has been unable to reproduce Baker's claims, which now appear discredited.

But Baker has led the field in recognising why men produce so many sperm – enough in one ejaculate theoretically to inseminate all the reproductive women in the world. The answer, with its origins in the evolutionary past, is sperm competition. When females mated with numerous males, sperm needed to compete to fertilise an egg – and the more sperm a male produced, the better his chances.

Plato in his *Republic* suggested that warriors should have the pick of the maidens because their sperm would improve the quality of the race. The 1946 Nobel Prize-winning American geneticist Hermann Muller held a similar view – but in favour of men of brain rather than muscle.

Muller long advocated establishing sperm banks in which donations from brilliant men would be stored. In the late 1970s, a Californian millionaire, Robert Graham, who believed 'retrograde humans' were gradually diluting the gene pool, followed Muller's advice, establishing his Repository for Germinal Choice.

Graham, famous as the inventor of the shatterproof spectacle lens, sold his company and focused on his vision, convincing three Nobel laureates to become donors. But elderly sperm, however eminent, proved poor for freezing and Graham cast the net wider to rising young scientists at universities, entrepreneurs, even Olympic gold-medal winners. Importantly, donors had to have an IQ of around 180 (there are estimated to be only about twenty people in Britain with this level of intelligence).

Graham was accused of being a eugenicist, but women – who had to be bright (Graham advertised in the Mensa magazine) and well off – came flocking. Donors' identities were kept secret, but such information as their weight, height, age, the colour of eyes, skin and hair and hereditary characteristics were given, from which women could make their choice, as can women seeking a donor today.[9]

That sperm, its 'fecundating' properties aside, influences female health and psychology is no new hypothesis – it goes back to the Ancient Greeks and Chinese and has been credited down the ages. In Restoration England adolescent girls, who frequently suffered from anaemia, were told they would be cured when marriage entitled them to regular infusions of it. The English birth-control pioneer Marie Stopes was saying much the same thing, affirming that 'the stimulating secretions which accompany man's semen' were highly beneficial when women absorbed them – which is why she disliked condoms (in fact, a wife in an 1872 divorce case complained that by using condoms her husband endangered her health). In the same period, Walter (*My Secret Life*) expressed the belief that 'spermatic lubrication is health-giving to a female' (making it sound, perhaps, like Lucozade). Tablets containing extract of semen were sold by a pharmacist in Chicago at the beginning of the twentieth century; and during the First World War, Harley Street doctors were prescribing 'male secretion treatment' for wives deprived of access to their usual source by the absence of their husbands in the trenches.

If the notion of semen as a kind of universal pick-me-up

seems absurd, a study conducted at the State University of New York in 2002 suggests otherwise. Researchers found that women who absorb semen vaginally are less depressed than those whose partners use condoms – and are twice as unlikely to attempt suicide as those who never have sex. The research team considered how often the women in the study had sex, the strength of their relationships, their personalities and whether or not they used oral contraceptives, and decided that all of these were irrelevant. Semen was the only factor in play. A more recent study tentatively concluded that swallowing semen while performing oral sex gives women some protection from breast cancer and pre-eclampsia, the dangerous high-blood-pressure disorder associated with pregnancy.

Not only does semen contain mood-altering hormones that make women happy – it's holistic.

SOME ARITHMETIC OF SEX

It may be true that Warren Beatty has had sex with in excess of 12,000 women; that the French writer Georges Simenon had sex with over 10,000 (most of whom he had to pay, mind); that the Italian dictator Mussolini had sex with a different woman every day for fourteen years. The rich and famous have a history of sexual excess. Once, such penis-possessors were called philanderers; today they are likely to be labelled sexual addicts and absolved for their transgressions by a whole new industry that has sprung up to treat them. Is there really such a thing as sexual addiction, analogous to addiction to alcohol or drugs? Or – moral issues aside – is having intercourse with multiple women only doing what comes naturally and would be done by most other penis-possessors given fame and riches (good looks possibly optional)?

Changing sexual mores mean that the average penis-possessor today is likely to have more sexual encounters than the average penis-possessor of earlier generations. A US university study of international promiscuity conducted in 2008 covered one-night stands, 'short-term matings' and longer relationships to assess what evolutionary psychologists term 'sociosexuality' – a measure of how sexually liberated people are in thought and behaviour. In this study the extrapolated information suggested that in the Western world the Finns were likely to have the most sexual partners over a lifetime – fifty-one; the British were placed eleventh with forty. But the findings were a *projection*. 'Running-total' data collected between 1999 and 2002 for the US National Center for Health Statistics found that up to that time in their lives a third of American penis-possessors had had fifteen partners, with the median being seven. A detailed survey in England in 2002 found that the average penis-possessor in London also claimed fifteen, with those outside London claiming twelve, both figures rising year on year. True or false? According to yet another research project these figures could possibly be halved; penis-possessors lie, even to themselves (as do non-penis-possessors, but in the opposite direction – they underestimate their tally in case they're stigmatised as promiscuous).

A penis-possessor's weekly tally of orgasms, however achieved, depends largely on age but is highly individual. Beyond the masturbatory teens when the desire for outlet may be every day or even several times a day, the weekly average over active adulthood – an indeterminate span – is two or three, a constant across cultures and centuries. (As in all things there are exceptions: Kinsey found a man who averaged thirty-three orgasms a week for thirty years.)

In marriage or relationships some penis-possessors, up to about the age of thirty, achieve three or perhaps four couplings a night,[10] but the average over active adulthood is twice a week, though the Durex global survey, conducted since 2006, suggests a global figure of 127 bouts of sex a year. Yet a breakdown by individual nationality delivers fluctuating results, with the

Hungarians, Bulgarians and Russians leading on 150, and the Swedes (102), Malays (100) and Singaporeans (96) trailing the field. The Japanese appear to be the least satisfied with their sex lives.

How often a penis-possessor has coital orgasm on any one occasion is fairly predictable: once or twice is the historical norm.

But while many never desire more than a single orgasm at any age, two is not enough for others, three is within the norm if hardly on the everyday scale, and anything beyond is exceptional. In his London journal James Boswell noted that in an encounter with 'Louisa' in 1763, he was 'fairly lost in supreme rapture' five times. In recent decades British tabloid newspapers have thought that Sir Ralph Halpern, the first British executive to earn £1 million, and inventor Sir Clive Sinclair deserving of banner headlines when young mistresses revealed them to be five-times-a-night men, despite both then being in their fifties; more banner headlines followed when an English lap dancer claimed that the Brazilian footballer Ronaldinho had scored (naturally) with her eight times in one evening. Only eight? The French novelist Victor Hugo, sexually hyperactive throughout his long life, said he made love to his wife nine times on their wedding night. Nine? According to anthropologists, ten times a night isn't out of the ordinary for the polygamous Chaggi of Tanzania. Ten? The sexually voracious actress Mae West, who had good-humoured contempt for the prowess of most men, was unstinting in her praise for a Frenchman named Dinjo who, she said, on one encounter coupled with her twenty-six times – a hard man good to find, indeed.

According to the World Health Organisation, there are an estimated 100 million acts of intercourse every day.

The rest is silence

During his laboratory investigations into human sexuality, the sexologist William Masters placed a finger in men's rectums during intercourse to assess how violently their prostate spasmed on ejaculation. In men over sixty he could detect nothing. It was, as it were, as if a mute had been clamped over the bridge of a stringed instrument: the strings still played but the resonance was deadened.

The decline in sexual prowess is relentless. From middle age onwards (whenever that may be) the erection that in his twenties sprang up without command now needs to be coaxed to attention, and it isn't what it was when it gets there; just as old age brings loss of muscle and height, so too it reduces the penis's dimensions: shoulders fully braced, the 6-inch erection of yesteryear is an inch or more shorter now. And where once a man could maintain erection for an hour during continuous sexual arousal before climax, he now finds it subsiding in less and less time; at sixty, according to Alfred Kinsey, six or seven minutes is all he, and his penis, can manage. There is also a progressive lessening of desire for intercourse; once a week, once a month, once now and again.

There can be compensation in a penis-possessor's sexual drive becoming less driven: in his fifties, when the urge is upon him, he's able to hold back in a way that was once beyond him; if he has learnt anything from life, he can give more consideration to his partner and gain satisfaction from this and from his greater control. In life generally, he can be less driven too – since the age of about thirty, testosterone, the rocket fuel of his sex drive but also of his aggression, has been falling by a percentage point a year. 'Twenty years ago, ten years ago, his dick would have been driving the car,' wrote Howard Jacobson about middle-aged Frank in his novel No More Mister Nice Guy. 'The great consolation of being fifty, for all your other organs, is that they finally get to sit behind the wheel.'

But the wheel of time turns, inexorably. True rigidity becomes a distant memory; the refractory period of sexual indifference after climax increases; the days of coming are going. Sexually speaking, men drop by the wayside. By sixty-five half of all men are, to use a sporting metaphor, out of the game; as are virtually all ten years later, without resort to chemical kick-starting.

South Sea islanders talk of a man's release from sexuality being like a boat that enters the tranquillity of the lagoon from the turbulence of the open sea. Some men might relate to this, regret even tinged with humour. (When he was seventy-eight, Winston Churchill, about to speak at a public meeting and passed a note by an aide telling him that his flies were unbuttoned, scrawled back: 'Dead birds do not fall out of nests.') Yet most are unlikely to admit to themselves that the lunchbox of youth has finally become the cocoa mug of old age. They, and their penis, like out-of-work actors, are only 'resting'.

Even in his dotage the possessor is likely still to consider himself a sexual being. 'The old man,' wrote the Italian poet Giacomo Leopardi, 'in the privacy of his thoughts, though he may protest the opposite, never stops believing that, through some singular exception of the universal rule, he can in some unknown and inexplicable way still make an impression on women.' In extreme cases this is nothing less than penile dementia. But however old he is, a man looks, oh he never stops looking: a pert breast, a pneumatic buttock, a face that makes him think: 'There was a time . . . ', or even, 'Even now, if . . . '; and if, and if . . .

Penis and possessor may no longer ask much of each other, or, if penis-possession finally becomes vacant possession, anything at all. But the prick of the brain, never mind its location or how neurologically defined, remains questing, even to the end of sentience.

PART FOUR NOTES

1. The indigenous San people of southern Africa, often referred to
    as 'Bushmen', are the oldest culture on the planet with a history
    dating back at least 70,000 years – and are known for their curious
    anatomy. The labia minora, or inner lips of the woman's vagina, are
    greatly enlarged, hanging down five or six inches like a small apron –
    and the males have a permanent semi-erection in the flaccid state. It
    was once speculated that San males were the last humans to have an
    os penis, but the condition results from a unique circular ligament in
    the rectus muscle.

2. A quirk of language is that fellation in modern times is invariably a
    'blow job', which suggests the very opposite of what the act involves.
       The derivation of the term isn't clear. As a euphemism for
    ejaculation, one supposition is that it's used in the sense of a kettle
    coming to the boil, or a whale surfacing: blowing off. Both seem
    fanciful. Far more persuasive is that the term comes from the black
    jazz scene of the late 1940s and '50s, meaning to play an instrument.
       The other modern alternative, 'giving head', needs no
    clarification.

3. For centuries a common method of contraception, anal sex is an
    increasing heterosexual recreational activity. Sixty years ago, Kinsey
    found that about one man in ten had tried it at least once, but he had
    little data, and none on women. In the 1990s a US national survey
    estimated that one man in five had tried anal sex – a figure that had
    doubled in 2005.
       One in three women admits to having experienced anal
    penetration but, according to a French study, only a third of them
    found it pleasurable.
       The prevalence of anal sex varies across the world with South
    Korea reporting the lowest figure and America the highest, where
    one in ten couples are believed to practise it regularly.

4. Hundreds of expressions for intercourse, as opposed to specific words,
    have or do exist, some of them highly personal between couples

(in Proust's *Remembrance of Things Past*, Swann begs Odette 'to do a camellia'). In the centuries when men got about on horses, for example, they talked of 'stabling the steed'; less elegantly now they talk about 'hiding the sausage', for which colloquialism thank the Australians.

5. For some men a partner's finger touching their prostate via their rectum before or during ejaculation is exquisitely pleasurable. For most men the spot is so sensitive that having it touched is painful, even stopping them from doing what they're doing. Many men, undergoing a medical rectal examination of their prostate, experience the embarrassment of becoming erect.

6. Writers of high-grade fiction strive to capture the intricacy of sexual activity, its psychological complexity, and to do it with originality. But on occasion the results are earnestly risible – which prompted the *Literary Review* magazine in 1993 to introduce its annual Bad Sex Award for Fiction. Every year the international competition is, to coin a cliché, stiff.

    Over the years, predictably, fictional efforts to portray the penis in varying states have garnered their share of *Review* nominations. Norman Mailer in his last novel (*Castle in the Forest*) compared an unerect penis to 'a soft coil of excrement'. John Updike wrote in *Brazil* of a character experiencing erection feeling 'his cashew become a banana, and then a rippled yam', and in *Seek My Face* wrote of an erection being like 'that Marisol Masterpiece with the cigarette lighter' (?). Rowan Somerville (*The Shape of Her*) became the 2010 winner for a description of penetration: 'Like a lepidopterist mounting a tough-skinned insect with a too blunt pin he screwed himself into her.' Perhaps Sarah Duncan was on to something saying in an article about the Bad Sex Award that her fellow fictive practitioners might be advised to eschew detailing what body parts go where and an explicit vocabulary; 'there isn't a single word for penis', she wrote, 'that doesn't sound daft. Dick, cock, willie, member, etc. They make me giggle.' (She has avoided appearing in the *Review*'s lists.)

7. Serious writers find themselves nominated by the *Literary Review* for its Bad Sex Award because their mazes of metaphors and similes

often make sex appear to have more to do with the natural elements than anything experienced between the sheets.

> Then I felt him beginning to move inside me again . . .
> Heaven or Hell . . . It was Heaven. I was the earth, the
> mountains, the tigers, the rivers that flowed into the lakes,
> the lake that became the sea. (*Eleven Minutes*, Paulo Coelho)

> As if struck by a sacred bolt of lightning . . . the world, the
> seagulls, the taste of salt, the hard earth, the smell of the sea,
> the clouds, all disappeared, and in their place appeared a vast
> golden light, which grew and grew until it touched the most
> distant star in the galaxy. (*Brida*, Paulo Coelho)

The Brazilian Coelho is not the only exponent of labyrinthine prolixity, of which the above is a mere sample. This, for example:

> Almost in an instant his desire . . . to reach a climax stalls and
> gives way to a sort of sensitive physical alertness . . . as though he
> has been transformed into a delicate seismograph that intercepts
> and instantly deciphers her body's reactions, translating . . . into
> skilful, precise navigation, anticipating and cautiously avoiding
> every sandbank, steering clear of each underwater reef . . .
> (*Rhyming Life and Death*, Amos Oz)

And this:

> She took my body into hers, and every movement was an
> incantation . . . My body was her chariot, and she drove it into
> the sun. Her body was my river, and I became the sea. And the
> wailing moan that drove our lips together, at the end, was the
> world of hope and sorrow that ecstasy wrings from lovers as it
> floods their souls with bliss (*Shantaram*, Gregory David Roberts)

Writers from lower down the literary scale don't escape the *Review*'s attention.

> His towel fell away. Sebastian's erect member was so big I
> mistook it for some sort of monument in the centre of a

town. I almost started directing traffic around it (*To Love, Honour and Betray*, Kathy Lette)

Her hand is moving away from my knee and heading north
. . . And, like Sir Ranulph Fiennes, Pamela will not easily be discouraged . . . Ever northward moves her hand . . . And when she reaches the north pole, I think in wonder and terror . . . she will surely want to pitch her tent (*Rescue Me*, Christopher Hart)

Honey Mackintosh bobbed up and down between my legs, her big soft lips locked around my hootchee and, true to her Scottish roots, she sucked away like she was the last person left on earth to play the bagpipes on Robbie Burns' birthday (*The Sucker's Kiss*, Alan Parker)

Meanwhile, down in Vaginaland, Mr Condom's beginning to feel a bit iffy . . . and as he gets flung willy-nilly in and out of the pink tunnel. He starts getting friction burns, hanging onto Bobby's stiff penis for dear life, headbutting Georgie's cervix at 180 beats per minute (*Ten Storey Love Song*, Richard Milward)

Whether comic sex writing is bad sex writing is a moot point. After all, its intention is to make readers laugh – not what those at the superior end of the writing trade really have in mind.

8. The Aranda hunters and gatherers of central Australia are one of the few peoples today who appear to be ignorant of the fact that males are necessary for procreation. Others are the Trobrianders, inhabitants of a group of islands in the western Pacific, the Yapese, inhabitants of a large island in the Carolines, and the Kajaba, a Colombian society.

The Keeraki of New Guinea believe that anal insemination could occur among boys who submit to men during puberty rites. The Etora in Oceania believe that boys have no semen at birth but acquire it through oral insemination by older men. New Guinea males have regular sex right up to the end of pregnancy because they believe that repeated infusions of semen provide the material to build the foetus's body. In Rwanda, tradition is that since semen and milk are white, intercourse is encouraged during lactation.

A number of societies believe semen can be transferred to the mother's milk, poisoning the baby. In Equatorial Guinea the Fang believe that a boy will become impotent if milk from his mother's breast drips on his penis.

9. The first donor insemination was conducted in 1884 at the Jefferson Medical College, Philadelphia, a medical student's semen being used on the wife of a Quaker merchant (who chose not to tell her). In subsequent years a few doctors conducted the procedure discreetly. In Britain during the 1930s unnamed senior academics were sperm donors, fathering large numbers of children. In a six-year period an eminent London neuropsychologist regularly supplied samples to a discreet London infertility clinic, his 'hyper-fecund' sperm leading to as many as two hundred births. It wasn't until 1954 that the public at large became aware of artificial insemination – a year after the first pregnancy was achieved using frozen sperm. In 2010 a baby was born from semen that had been frozen for over twenty years.

Sperm donors must be fit, free of disease and their sperm count above average to give the best chance of pregnancy. Fresh donations have greater fecundity than frozen and produce higher pregnancy rates. Many countries have a serious shortage of donors, often because they don't allow anonymous donation, including Britain. The most sought-after sperm on the global market comes from Denmark – the Danes tend to be tall, highly educated and donate for altruistic motives.

Sperm agencies recruit contributors, usually via the Internet, supplying them with a collection kit and a courier service. Agencies are largely unregulated. Most women seeking a donor do so through strictly controlled sperm banks and fertility clinics – where men provide their samples while viewing erotic material.

And what of Robert Graham's supremacist genetic experiment? Some hundreds of children were born as a result of it. But after he died in 1997 at the age of ninety, his repository, sited in an underground bunker on his ranch near San Diego, no longer funded, closed down.

10. Patriarchal societies though they were, the Ancient Greeks and early Jews enshrined a man's marital duty in law. The Athenians decreed that citizen wives were entitled to sex three times a month. The Jews were more specific: labourers were to satisfy their wife to orgasm

twice a week; scholars only on Fridays; businessmen who travelled to other cities, once a week; and camel drivers (who travelled even more and whose work was arduous) once every thirty days. 'Men of leisure' were theoretically under obligation to satisfy their wives nightly, which seems to have set the bar a bit high.

EPILOGUE

As two

The penis is not just a body part – it is a determinant of identity and behaviour. Possession can lift the possessor to great heights or hurtle him to great depths; as in any long-term relationship there are good days and bad days.

The difference in libido, that tangled combination of heart and head and hormones, between men and women is vast. Men are programmed to hunt, to spread their seed, women to conserve and nurture, though that is a simplification. What is true is that penis-possession can allow men to depersonalise sex almost at will, if they have a will, and throughout adult life makes 'infinite desirelessness a strain to bring on' (*Intimacy*, Hanif Kureishi). Should men rove, or venture where they wished they had not, they're likely to blame their penis, not themselves, just as they do should they fail in their performance. It is as if, as the Californian sex therapist Barbara Keesling quipped, their penis is 'a stranger letting space in their underwear'. And a possessor almost always forgives his penis any transgressions; it's unlikely that he will admonish his appendage further than the hero of Paul Theroux's *My Secret History*, who 'often looked at [his] penis and thought: you *moron*'. How could it be otherwise?

Women have more to lose than men by submitting to the libido's lash, yet men don't have a monopoly on infidelity; humankind is only nominally monogamous. There is, of course, love, which may be for ever, or not. Women prefer sex to be linked to love. Men often aren't worried. Men succumb to love more quickly; women need time, just as they do in bed. And in bed the different paths that men and women want to follow to arousal and completion can cause all kinds of misunderstandings. Even at their closest the planets Mars and Venus are 128 million miles apart.

If sex is so difficult, if men and women are such an imperfect match, why does sex seem to dominate their lives? After all, their primate cousins (other than bonobo chimpanzees) couple only when the female is fertile (which in the gorilla's case is for six days every four years – even the dominant male in a group is lucky to have sex a few times a year). Humans, Jared Diamond points out, 'are bizarre in our near continuous practice of sex'.

And why? Because, as Diamond also points out, human sex isn't just a biological imperative. Despite all the trouble and travail it causes penis-possessors and non-penis-possessors alike, sex is fun. A world without sex, the American humorist Henry Louis Mencken thought,

> would be unbearably dull. It is the sex instinct which makes
> women seem beautiful, which they are once in a blue
> moon, and men seem wise and brave, which they never are
> at all. Throttle it, denaturalize it, take it away, and human
> existence would be reduced to the prosaic, laborious,
> boresome, imbecile level of life in an anthill.

Yet, increasingly, science is divorcing sex from reproduction. The *Journal of Australia*, an authority on in-vitro fertilisation, predicts that sexual intercourse will fade away as a means of procreation in favour of technology that can achieve genetic preferences and avoid genetic risks. Evolutionary biologist Robin Baker goes further, predicting that 'if sex no longer has any biological benefit but is only acting out ancient scenarios', its attraction could wane. Should that happen the penis would become less and less significant and in a thousand years could shrink to the gorilla-like dimensions from which it grew.

It's a dystopian view. Penis-possessors everywhere can only hope that their descendants fight evolutionary fate every inch of the way.

INDEX